Lecture Notes in Computer Science 12925

More information about this subseries at http://www.springer.com/series/7409

Matteo Golfarelli · Robert Wrembel ·
Gabriele Kotsis · A Min Tjoa ·
Ismail Khalil (Eds.)

Big Data Analytics and Knowledge Discovery

23rd International Conference, DaWaK 2021
Virtual Event, September 27–30, 2021
Proceedings

Editors
Matteo Golfarelli (iD)
University of Bologna
Bologna, Forli/Cesena, Italy

Robert Wrembel
Poznań University of Technology
Poznan, Poland

Gabriele Kotsis
Johannes Kepler University Linz
Linz, Austria

A Min Tjoa
TU Wien
Vienna, Austria

Ismail Khalil
Johannes Kepler University Linz
Linz, Austria

ISSN 0302-9743 ISSN 1611-3349 (electronic)
Lecture Notes in Computer Science
ISBN 978-3-030-86533-7 ISBN 978-3-030-86534-4 (eBook)
https://doi.org/10.1007/978-3-030-86534-4

LNCS Sublibrary: SL3 – Information Systems and Applications, incl. Internet/Web, and HCI

Preface

DaWaK was established in 1999 as an International Conference on Data Warehousing and Knowledge Discovery. It was run continuously under this name until its 16th edition (2014, Munich, Germany). In 2015 (Valencia, Spain) it was renamed to International Conference on Big Data Analytics and Knowledge Discovery (retaining the DaWaK acronym) to better reflect new research directions in the broad and dynamically developing area of data analytics. In 2021, the 23rd edition of DaWaK was held virtually during September 27–30, 2021.

Since the very beginning, the DaWaK conference has been a high-quality forum for researchers, practitioners, and developers in the field of data integration, data warehousing, data analytics, and recently big data analytics, in a broader sense. The main objectives of this event are to explore, disseminate, and exchange knowledge in these fields through scientific and industry talks. Big data analytics and knowledge discovery remain hot research areas for both academia and the software industry. They are continuously evolving, being fuelled by advances in hardware and software. Important research topics associated with these major areas include: data lakes (schema-free repositories of heterogeneous data); database conceptual, logical, and physical design; data integration (especially linking structured and semistructured data sources); big data management (mixing relational tables, text, and any types of files); query languages (SQL and beyond); scalable analytic algorithms; parallel systems (cloud, parallel database systems, Spark, MapReduce, HDFS); theoretical foundations for data engineering; and practical applications.

DaWaK 2021 attracted 71 papers (whose authors came from 17 countries), from which the Program Committee (PC) selected 12 full papers and 15 short papers, yielding an acceptance rate of 16% for the full paper category and 35% for both categories. Each paper was reviewed by at least three reviewers and in some cases up to four. Accepted papers cover a variety of research topics on both theoretical and practical aspects. The program includes the following topics: (1) data warehouse and materialized views maintenance, (2) system performance, (3) data quality, (4) advanced analytics, prediction techniques, and machine learning, (5) text processing and analytics, as well as (6) knowledge representation, revisited in the current technological contexts.

Thanks to the reputation of DaWaK, selected best papers of DaWaK 2021 were invited for publication in a special issue of the *Data & Knowledge Engineering* (*DKE*, Elsevier) journal. Therefore, the PC chairs would like to thank the *DKE* Editor in-Chief, Prof. Carson Woo, for his approval of the special issue.

Although the pandemic struck many aspects of daily life, DaWak 2021 succeeded in terms of the number of submissions and the quality of submitted papers. We would like to express our sincere gratitude to all Program Committee members and the external reviewers who reviewed the papers thoroughly and in a timely manner. Finally, we would like to thank the DEXA conference organizers for their support and guidance, especially, Prof. Ismail Khalil for having provided a great deal of assistance and for putting his experience at our disposal.

September 2021

Matteo Golfarelli
Robert Wrembel

Organization

Program Committee Chairs

Matteo Golfarelli	University of Bologna, Italy
Robert Wrembel	Poznan University of Technology, Poland

Steering Committee

Gabriele Kotsis	Johannes Kepler University Linz, Austria
A Min Tjoa	Vienna University of Technology, Austria
Robert Wille	Software Competence Center Hagenberg, Austria
Bernhard Moser	Software Competence Center Hagenberg, Austria
Ismail Khalil	Johannes Kepler University Linz, Austria

Program Committee

Alberto Abello	Universitat Politècnica de Catalunya, Spain
Toshiyuki Amagasa	University of Tsukuba, Japan
Amin Beheshti	Macquarie University, Australia
Amina Belhassena	Harbin Institute of Technology, China
Ladjel Bellatreche	LIAS, ENSMA, France
Soumia Benkrid	Ecole Nationale Supérieure d'Informatique, Algeria
Fadila Bentayeb	ERIC, Université Lumière Lyon 2, France
Jorge Bernardino	ISEC, Polytechnic Institute of Coimbra, Portugal
Vasudha Bhatnagar	University of Delhi, India
Besim Bilalli	Universitat Politècnica de Catalunya, Spain
Sandro Bimonte	Irstea, France
Pawel Boinski	Poznan University of Technology, Poland
Kamel Boukhalfa	University of Science and Technology Houari Boumediene, Algeria
Omar Boussaid	ERIC Laboratory, France
Michael Brenner	Leibniz Universität Hannover, Germany
Stephane Bressan	National University of Singapore, Singapore
Wellington Cabrera	Teradata Labs, USA
Joel Luis Carbonera	Federal University of Rio Grande do Sul, Brazil
Frans Coenen	University of Liverpool, UK
Isabelle Comyn-Wattiau	ESSEC Business School, France
Alfredo Cuzzocrea	ICAR-CNR and University of Calabria, Italy
Laurent D'Orazio	University Rennes, CNRS, IRISA, France
Jérôme Darmont	Université de Lyon, France
Soumyava Das	Teradata Labs, USA
Karen Davis	Miami University, USA

Ibrahim Dellal	LIAS, ENSMA, France
Alin Dobra	University of Florida, USA
Markus Endres	University of Passau, Germany
Leonidas Fegaras	University of Texas at Arlington, USA
Philippe Fournier-Viger	Harbin Institute of Technology, Shenzhen, China
Matteo Francia	University of Bologna, Italy
Filippo Furfaro	University of Calabria, Italy
Pedro Furtado	Universidade de Coimbra, Portugal
Luca Gagliardelli	University of Modena and Reggio Emilia, Italy
Enrico Gallinucci	University of Bologna, Italy
Johann Gamper	Free University of Bozen-Bolzano, Italy
Kazuo Goda	University of Tokyo, Japan
María Teresa Gómez-López	University of Seville, Spain
Anna Gorawska	Wroclaw University of Technology, Poland
Marcin Gorawski	Silesian University of Technology and Wroclaw University of Technology, Poland
Sven Groppe	University of Lübeck, Germany
Hyoil Han	Illinois State University, USA
Frank Höppner	Ostfalia University of Applied Sciences, Germany
Stephane Jean	LISI, ENSMA, and University of Poitiers, France
Selma Khouri	Ecole Nationale Supérieure d'Informatique, Algeria
Jens Lechtenbörger	University of Münster, Germany
Wookey Lee	Inha University, South Korea
Young-Koo Lee	Kyung Hee University, South Korea
Wolfgang Lehner	TU Dresden, Germany
Carson Leung	University of Manitoba, Canada
Sebastian Link	University of Auckland, New Zealand
Sofian Maabout	LaBRI, University of Bordeaux, France
Patrick Marcel	Université de Tours, LIFAT, France
Amin Mesmoudi	LIAS, Université de Poitiers, Franc
Jun Miyazaki	Tokyo Institute of Technology, Japan
Anirban Mondal	University of Tokyo, Japan
Rim Moussa	ENICarthage, Tunisia
Kjetil Nørvåg	Norwegian University of Science and Technology, Norway
Boris Novikov	National Research University Higher School of Economics, Russia
Makoto Onizuka	Osaka University, Japan
Carlos Ordonez	University of Houston, USA
Jesus Peral	University of Alicante, Spain
Uday Kiran Rage	University of Tokyo, Japan
Praveen Rao	University of Missouri, USA
Franck Ravat	IRIT, Université de Toulouse, France
Oscar Romero	Universitat Politècnica de Catalunya, Spain
Anisa Rula	University of Brescia, Italy
Keun Ho Ryu	Chungbuk National University, South Korea

Ilya Safro	University of Delaware, USA
Kai-Uwe Sattler	Ilmenau University of Technology, Germany
Monica Scannapieco	Italian National Institute of Statistics, Italy
Alkis Simitsis	HP Labs, USA
Emanuele Storti	Università Politecnica delle Marche, Italy
Olivier Teste	IRIT, France
Dimitri Theodoratos	New Jersey Institute of Technology, USA
Maik Thiele	TU Dresden, Germany
Panos Vassiliadis	University of Ioannina, Greece
Lena Wiese	Goethe University Frankfurt, Germany
Szymon Wilk	Poznan University of Technology, Poland
Sadok Ben Yahia	Université de Tunis El Manar, Tunisia
Haruo Yokota	Tokyo Institute of Technology, Japan
Hwan-Seung Yong	Ewha Womans University, South Korea
Yinuo Zhang	Teradata Labs, USA
Yongjun Zhu	Sungkyunkwan University, South Korea

Additional Reviewers

Giuseppe Attanasio
Pronaya Prosun Das
Jorge Galicia
Rediana Koçi
Michael Lan
Dihia Lanasri
Vivien Leonard
M. Saqib Nawaz
Uchechukwu Njoku
Mourad Nouioua
Moreno La Quatra
Shivika Prasanna
Andrea Tagarelli
Luca Virgili
Xiaoying Wu
Arun George Zachariah

Organizers

Contents

Advanced Analytics

Machine Learning and Deep Learning

Data Warehouse Processes and Maintenance

Machine Learning and Analtyics

Performance

Bounding Box Representation of Co-location Instances for L_∞ Induced Distance Measure

Witold Andrzejewski[✉] and Pawel Boinski[✉]

Institute of Computing Science, Poznan University of Technology, Piotrowo 2,
60-965 Poznan, Poland
witold.andrzejewski,pawel.boinski}@put.poznan.pl

Abstract. In this paper, we investigate the efficiency of Co-location Pattern Mining (CPM). In popular methods for CPM, the most time-consuming step consists of identifying of pattern instances, which are required to calculate the potential interestingness of the pattern. We tackle this problem and provide an instance identification method that has lower complexity than the state-of-the-art approach: (1) we introduce a new representation of co-location instances based on bounding boxes, (2) we formulate and prove several theorems regarding such a representation that can improve instances identification step, (3) we provide a novel algorithm utilizing the aforementioned theorems and analyze its complexity. Finally, we experimentally demonstrate the efficiency of the proposed solution.

Keywords: Co-location · Bounding box · Data mining

1 Introduction

The problem of CPM was defined as searching for sets of spatial features whose instances are often in their proximity [1]. A spatial feature can be regarded as a type of object, while a spatial feature instance is an object of that type. For example, assume that a railroad station is a spatial feature. In such a case, King's Cross station in London is an instance of that spatial feature. Knowledge about co-locations can be very useful in domains like epidemiology [2], geology [3], health care and environmental sciences [4] etc.

Over the years, various algorithms for CPM have been developed. Many of them were based on methods known from Frequent Itemset Mining (FIM) [5], which is not surprising as CPM has roots in FIM. In both cases, we are looking for elements that often occur together, either among transactions in the famous shopping cart problem or close to each other in the case of co-locations. However, in FIM, the number of analyzed element types (e.g. products) is very high, while the number of spatial features in CPM is significantly limited. The main difference is in the pattern instances identification. In contrast to FIM, in

© Springer Nature Switzerland AG 2021
M. Golfarelli et al. (Eds.): DaWaK 2021, LNCS 12925, pp. 3–14, 2021.
https://doi.org/10.1007/978-3-030-86534-4_1

which transactions are well defined, the continuity of space in CPM makes the identification of co-location instances not trivial. Moreover, in dense datasets, the number of such instances can be tremendous. While FIM and CPM can share similar methods for generating candidates for patterns or even concepts of interestingness measures, the identification of candidate/pattern instances is very different and has become the main topic of many researches. Historically, such identification methods started from a spatial join, through a partial join, and ended with a join-less approach.

In this paper, we address the problem of candidate/co-location instance identification by introducing a new idea for co-location instance representation based on bounding boxes. We prove that such a representation can be used for efficient identification of co-location instances and we provide an algorithm for that purpose. We also present a discussion regarding a reduced computational complexity of that approach in comparison to the state-of-the-art solution.

2 Related Work

The formal definition of CPM was introduced in [1]. In the same work, the authors proposed a general algorithm for co-location pattern discovery based on the very popular (at that time) approach to mining patterns. The general algorithm consists of two steps: (1) generating co-location candidates and (2) filtering candidates w.r.t. interestingness measure. Candidates that meet the assumed condition become co-location patterns. The authors proposed a co-location prevalence measure to determine the interestingness of candidates. The prevalence measure has the antimonotonicity property, which allows reducing the number of generated candidates using the iterative Apriori-based method [5]. Starting from initial element co-locations (size $k = 1$), in each iteration, size $k+1$ candidates are generated and tested w.r.t. the prevalence measure. These two steps are performed as long as it is possible to generate new candidates.

The first algorithm, utilizing mentioned approach, applied a computationally expensive spatial join to obtain size k candidate instances from size $k - 1$ instances of discovered patterns. The spatial join-based method is limited only to sparse or small datasets. In subsequent studies, alternative methods for instance identification have been proposed. In [6], the authors introduced the algorithm that can divide input dataset into artificial transactions, however it requires auxiliary computations to trace inter-transactions neighbors. Yoo et al. proposed a joinless method based on the fast generation of a superset of instances [7]. Such superset must be verified in an additional step that checks the correctness of each instance. Following this path, Wang et al. proposed an instance generation method similar to the joinless algorithm but producing only correct results [8]. For this purpose, the authors utilized the concept of star-type neighborhoods in the form of a new structure called an improved Co-location Pattern Instance tree (iCPI-tree). Practical implementation of this state-of-the-art approach proved to be very efficient and has been adopted for massively parallel execution using modern GPUs [9].

In this paper, we adopt the iCPI-tree by introducing a new concept of co-location instances discovery that utilizes the idea of bounding boxes [10] which are widely used, e.g. in spatial indexing [11]. We start with recalling basic definitions related to CPM in Sect. 3.1. Next, in Sect. 3.2, we provide new definitions and proofs regarding the novel algorithm presented in Sect. 4. In Sect. 5, we show the results of performed experiments. The paper ends with a summary in Sect. 6.

3 Definitions

3.1 Common Definitions

Definition 1. *Let f be a* **spatial feature**, *i.e. characteristics of space in a given position. We say that an object o is an* **instance of the feature** f, *if o has a property f and is described by a position and a unique identifier. A set of all features is denoted as F. We assume that features are ordered w.r.t. total order $\leq_F$ defined for set F. Let S be a set of objects with spatial features in F. Any such set is called a* **spatial dataset**.

We use the following notations. If the feature of an object is not important, the object is denoted as o, optionally with a subscript, e.g. o_i, o_1 etc. An object with the feature f is denoted as o^f. Both of these notations may be combined.

Definition 2. *Object's location in space is called a* **position**. *Let K be the set of all possible positions in space. In this paper, we consider space for which $K = \mathbb{R}^m$. Hence, a position is an m dimensional vector of real values called* **coordinates**. *Given a position $k \in K$, an i-th coordinate ($i = 1, \ldots, m$) is denoted as k_i. A function p (called a* **position function**) *is a function that assigns a position to an object.*

Definition 3. *A* **distance function** $dist : K \times K \to \mathbb{R}^+ \cup \{0\}$ *is any distance metric that assigns to every pair of positions in K the distance between them.*

In this paper we focus on a distance function induced by L_∞ norm. We define such distance function as $dist_\infty(a, b) = L_\infty(a - b) = max_{i=1,\ldots m}\{|a_i - b_i|\}$.

Definition 4. *Let d be any value in $\mathbb{R}^+ \cup \{0\}$. The relation defined as $R(d) = \{(o_i, o_j) : o_i, o_j \in S \wedge dist(p(o_i), p(o_j)) \leq d\}$ is called a* **neighborhood relation** *and the d value is called a* **maximal distance threshold**. *Any two objects o_i and o_j such that $(o_i, o_j) \in R(d)$ are called a* **neighbor pair**. *The object o_i is called the* **neighbor** *of the object o_j and vice versa.*

Definition 5. *Given a spatial object $o_i \in S$ whose spatial feature is $f_i \in F$, the* **neighborhood** *of o_i is defined as a set of spatial objects: $T_{o_i} = \{o_j \in S : o_j = o_i \vee (f_i < f_j \wedge (o_i, o_j) \in R(d))\}$ where $f_j \in F$ is the feature of o_j and R is a neighborhood relation. A set of all objects in the neighborhood T_{o_i} with the feature f $T_{o_i}^f = \{o^f : o^f \in T_{o_i}\}$ is called the* **limited neighborhood**.

Definition 6. *A **co-location pattern** (co-location in short) c is a non-empty subset of spatial features $c \subseteq F$. We say that set $i^c \subseteq S$ is an **instance of co-location** c if: (1) i^c contains instances of all spatial features of c and no proper subset of i^c meets this requirement, (2) all pairs of objects are neighbors, i.e. $\forall_{o_i, o_j \in i^c} (o_i, o_j) \in R(d)$.*

Definition 7. *The **participation ratio** $Pr(c, f_i)$ of a feature f_i in a co-location $c = \{f_1, f_2, \ldots, f_k\}$ is a fraction of objects representing feature f_i in the neighborhood of instances of co-location $c \backslash \{f_i\}$, i.e. $Pr(c, f_i) = \frac{|distinct\ instances\ of\ f_i\ in\ instances\ of\ c|}{|instances\ of\ f_i|}$.*

Definition 8. *The **participation index** $Pi(c)$ of a co-location $c = \{f_1, f_2, \ldots, f_k\}$ is defined as $Pi(c) = \min_{f_i \in c} \{Pr(c, f_i)\}$. Participation index is called a **prevalence** of a co-location pattern.*

Problem 1. Given a dataset S, distance function $dist$, as well as d and $minprev$ thresholds, the **prevalent spatial co-location pattern mining** problem is finding a set P of all spatial co-location patterns c such that $Pi(c) \geq minprev$.

3.2   Co-location Instance Representation

In order to introduce the new idea for the co-location instance representation, we need to introduce several definitions and prove several lemmas and theorems.

Definition 9. *Given any m dimensional space K, a **bounding box** bb is a set of positions in that space contained in an axis-aligned cuboid. We represent such a bounding box as a list of closed intervals, one for each dimension, i.e. $([a_1, b_1], [a_2, b_2], \ldots [a_m, b_m])$. An interval of bounding box bb for dimension j is denoted as $int_j(bb)$.*

Definition 10. *Given a feature instance o (located in m dimensional space K) and distance d, we define a **feature instance bounding box** as follows. Let $p(o) = [q_1, q_2, \ldots, q_m]$ represent the feature instance position. Feature instance bounding box, denoted $M(o)$, is defined as a list of intervals $\left[q_i - \frac{d}{2}, q_i + \frac{d}{2}\right]$ for $i = 1, \ldots, m$. For brevity, we will use an acronym OBB instead of "feature instance bounding box".*

Theorem 1. *Given the distance function $dist_\infty$ and threshold d, two objects o_1 and o_2 are neighbors if, and only if, the intersection of their bounding boxes is nonempty, i.e. $M(o_1) \cap M(o_2) \neq \emptyset \iff dist_\infty(o_1, o_2) \leq d$.*

Proof. We separately prove two directions of the implication. Let o_1 and o_2 be two objects with positions $p(o_1) = [q_1^1, q_2^1, \ldots, q_m^1]$ and $p(o_2) = [q_1^2, q_2^2, \ldots, q_m^2]$.
a) $M(o_1) \cap M(o_2) \neq \emptyset \implies dist_\infty(o_1, o_2) \leq d$.
   Since the objects' bounding boxes intersect, then one of the following inequalities is true at each dimension $j = 1, 2, \ldots, m$: (1) $q_j^1 + \frac{d}{2} \geq q_j^2 - \frac{d}{2}$, (2) $q_j^2 + \frac{d}{2} \geq$

$q_j^1 - \frac{d}{2}$. By reordering and simplifying the terms in these equations we get: (1) $q_j^2 - q_j^1 \leq d$, (2) $q_j^1 - q_j^2 \leq d$. Hence, $|q_j^1 - q_j^2| \leq d$. Because this is true for every dimension j, then $max_{j \in 1,\ldots,m}(|q_j^1 - q_j^2|) \leq d$. Since $max_{j \in 1,\ldots,m}(|q_j^1 - q_j^2|) = dist_\infty(o_1, o_2)$ this result finalizes the proof.

b) $M(o_1) \cap M(o_2) \neq \emptyset \impliedby dist_\infty(o_1, o_2) \leq d$

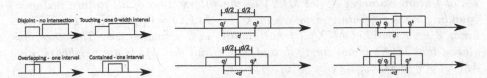

Fig. 1. Types of interval intersections

Fig. 2. Possible intersections of intervals for outermost objects

Fig. 3. Intersections with intervals for non outermost objects

We assume that $dist_\infty(o_1, o_2) \leq d$. Hence, $max_{j \in 1,\ldots,m}(|q_j^1 - q_j^2|) \leq d$. For this to be true, every term $|q_j^1 - q_j^2|$ should be less than or equal to d, i.e. $\forall_{j \in 1,\ldots,m}|q_j^1 - q_j^2| \leq d$. Therefore, for each dimension $j = 1, 2, \ldots, m$ one of the following inequalities should be true: (1) $q_j^2 - q_j^1 \leq d$, (2) $q_j^1 - q_j^2 \leq d$. By splitting d into a sum of two $\frac{d}{2}$ and reordering the terms we obtain the following inequalities: (1) $q_j^1 + \frac{d}{2} \geq q_j^2 - \frac{d}{2}$, (2) $q_j^2 + \frac{d}{2} \geq q_j^1 - \frac{d}{2}$. Since these inequalities, if true for all dimensions, represent situation in which two bounding boxes intersect, this finalizes the proof.

Theorem 2. *Given any set of bounding boxes BB, their intersection (if not empty) is also a bounding box. Formally, $\bigcap_{bb_i \in BB} bb_i$ is a bounding box (if not empty).*

Proof. For the intersection of a set of bounding boxes to be not empty, their intervals for each of the dimensions must intersect. Hence, it is sufficient to show that if all the intervals for an arbitrary dimension intersect, the result is a single closed interval.

For the purpose of the discussion, let us assign any order to the bounding boxes in the BB set, i.e. $BB = \{bb_1, bb_2, \ldots, bb_{|BB|}\}$. Since set intersection is a binary associative, commutative and closed operator, the order of intersections performed in $\bigcap_{bb_i \in BB} bb_i$ is arbitrary. Assume that bb_1 and bb_2 are intersected first, the result is intersected with bb_3 and so on, i.e. $\bigcap_{bb_i \in BB} bb_i = (((bb_1 \cap bb_2) \cap bb_3) \cap \ldots) \cap bb_{|BB|}$. This operation is performed for each dimension separately, thus for each dimension $j = 1, \ldots, m$, the following intersections are performed:

$$(((int_j(bb_1) \cap int_j(bb_2)) \cap int_j(bb_3)) \cap \ldots) \cap int_j(bb_{|BB|}) \qquad (1)$$

When two intervals intersect, the result can only be empty or be composed of only a single interval (see Fig. 1). Hence, after each intersection performed in Eq. 1 we obtain at most a single interval as a result. Thus, if the result is not empty for every dimension $j = 1, \ldots, m$ then all the bounding boxes in BB have an intersection, which is also a bounding box.

Theorem 3. *Given the distance function $dist_\infty$, distance threshold d and a set of feature instances X, let $M(X)$ be the intersection of all feature instance bounding boxes for feature instances in X, i.e. $M(X) = \bigcap_{o \in X} M(o)$. From Theorem 2 we know that $M(X)$ (if not empty) is a bounding box. This theorem states, that $M(X)$ is not empty if, and only if, all the objects are neighbors, i.e. $M(X) \neq \emptyset \iff \forall_{o_1, o_2 \in X} dist_\infty(o_1, o_2) \leq d$.*

Proof. We separately prove two directions of the implication.

a) $M(X) \neq \emptyset \impliedby \forall_{o_1, o_2 \in X} dist_\infty(o_1, o_2) \leq d$

Let X be a set of objects o_i, $i = 1, \ldots, |X|$. Each object is located at $p(o_i) = \left[q_1^i, q_2^i, \ldots, q_m^i\right]$. Each objects' bounding box is defined by a set of intervals $int_j(o_i) = \left[q_j^i - \frac{d}{2}, q_j^i + \frac{d}{2}\right]$ for $j = 1, \ldots, m$. In order for the bounding box $M(X)$ to be not empty, for each dimension these intervals must intersect for all objects. Thus, in the following discussion we show that if $\forall_{o_i, o_k \in X} dist_\infty(o_i, o_k) \leq d$, these intervals intersect in all dimensions $j = 1, 2, \ldots, m$.

Let us mark all the objects' locations on a dimension j axis. Assume o_i and o_k are the two outer (most distant) objects, where $q_j^i \leq q_j^k$. Since $dist_\infty(o_i, o_k) \leq d$, the distance between all objects' pairs in every dimension j must be less or equal to d. Hence, the intervals $\left[q_j^i - \frac{d}{2}, q_j^i + \frac{d}{2}\right]$ and $\left[q_j^k - \frac{d}{2}, q_j^k + \frac{d}{2}\right]$ will always intersect. In the worst case (if the distance between o_j and o_k on dimension i is equal to d), the intersection will contain only the boundary exactly in the middle between the objects, i.e. position $q_i^j + \frac{d}{2} = q_k^j - \frac{d}{2}$ (see Fig. 2). Since all the other objects are located between o_i and o_k, they are closer to the middle of the distance between o_j and o_k and thus, their bounding box interval for dimension j will contain that point as well (see Fig. 3).

Since the above discussion is valid for any dimension $j = 1, \ldots, m$, this shows that if $\forall_{o_1, o_2 \in i^c} dist_\infty(o_1, o_2) \leq d$ then $M(X) \neq \emptyset$, which finalizes the proof.

b) $M(X) \neq \emptyset \implies \forall_{o_1, o_2 \in X} dist_\infty(o_1, o_2) \leq d$

If the bounding box $M(X)$ is not empty, all pairwise intersections of feature instance bounding boxes of feature instances in X are not empty as well. In such a case, Theorem 1 states that all pairs of objects in X are neighbors.

Definition 11. *Given co-location instance i^c let $M(i^c) = \bigcap_{o \in i^c} M(o)$. From Theorem 2 we know that $M(i^c)$ is a bounding box. Hence, the set $M(i^c)$ is called a **co-location instance center bounding box**. For brevity, we will use an acronym CBB.*

Theorem 3 provides interesting properties of CBBs. Since, by definition, all the objects in i^c are neighbors, then $M(i^c)$ is not empty. However, a more important consequence of this theorem is an observation that given any set of feature instances with distinct features X, X can only be co-location instance if $M(X)$ is not empty. This fact leads to optimizations described in the following section.

4 Instance Identification

In many prevalent co-location pattern mining algorithms, the most time-consuming step is identifying candidate co-location pattern instances. This section demonstrates that (as long as $dist_\infty$ distance function is used) it is possible to reduce the complexity of this step by taking advantage of the properties of co-location instance center bounding boxes introduced in Sect. 3.2.

One of the fastest, state-of-the-art algorithms for mining prevalent co-location patterns uses an iCPI-tree structure for the identification of candidate co-location instances [8]. The iCPI-tree is a structure that allows constant retrieval time of limited neighborhoods for given feature instances o and the neighbor features f, i.e. T_o^f. The candidate instance identification algorithm requires the co-location pattern candidate c, a list of instances $I^{c'}$ of the candidate's prefix c' ($|c| - 1$ smallest features in the order $\leq_F$) obtained in previous iteration of the main algorithm, and the iCPI-tree structure. The idea for the algorithm is as follows. Given the candidate c, we split it into its prefix c' and an extending feature $\hat{c}$ (largest feature in the order $\leq_F$). Given any instance $i^{c'} \in I^{c'}$, we know that all the feature instances in it are neighbors. Thus, to find an instance of the candidate c, i.e. i^c, we only need to find an instance of the candidate's extending feature $\hat{c}$, i.e. $o^{\hat{c}}$, such that it is a neighbor of all feature instances in $i^{c'}$. To find all of such instances, it is sufficient to find an intersection of limited neighborhoods of all objects in $i^{c'}$ with feature $\hat{c}$, formally $\bigcap_{o \in i^{c'}} T_o^{\hat{c}}$. We provide the sketch of this algorithm in Algorithm 1.

Let us estimate the complexity of this algorithm. The complexity of multiple set intersections strongly depends on the data and the data structures used. Assuming a sorted list or a hash table representation, intersection requires the number of steps in the order of the number of elements in both of the sets (sorted list) or one of the sets (hash table). Since limited neighborhoods are intersected, we assume that the complexity of the intersection operation is $O(|T_o^{\hat{c}}|)$. In order to find common neighbors of all feature instances in $i^{c'}$, we need to perform $|i^{c'}| - 1$ intersections. Thus, the inner loop (lines 6–8) requires $O(|i^{c'}||T_o^{\hat{c}}|)$ operations. The second inner loop has as many iterations as there are results in the first inner loop. Hence, its complexity is the same or lower. Finally, the outer loop is repeated $|I^{c'}|$ times. Thus, final complexity order of the algorithm is $O(|I^{c'}||i^{c'}||T_o^{\hat{c}}|)$.

The complexity of the candidate instance identification algorithm can be decreased using the theorems and definitions introduced in Sect. 3.2. The basic idea is as follows. Assume a candidate c. Let $M(i^{c'})$ be a CBB corresponding to an instance of the prefix c' of the candidate. Let $o^{\hat{c}}$ be an object with extending feature $\hat{c}$ and $M(o^{\hat{c}})$ be its OBB. Let $X = M(o^{\hat{c}}) \cap M(i^{c'}) = M(o^{\hat{c}}) \cap \bigcap_{o \in i^{c'}} M(o)$. Theorem 2 states that X (if not empty) is a bounding box, while Theorem 3 states that if X is not empty, then all the objects in $i^{c'} \cup \{o^{\hat{c}}\}$ are neighbors. Note that, in such a case, $i^{c'} \cup \{o^{\hat{c}}\} = i^c$ is an instance of the candidate c while $X = M(i^c)$ is its CBB. The most important observation is that we can instantly determine whether one feature instance is a neighbor of all feature instances

---

Algorithm 1. Instance identification in the iCPI-tree algorithm

input:
    a co-location candidate c
    a list of instances of the candidate's prefix $I^{c'}$
    an iCPI-tree for quick retrieval of limited neighborhoods T_o^f
output:
    a list of the candidate's instances I^c
1: $I^c \leftarrow \emptyset$
2: **for** $k \in 1, 2, \ldots, |I^{c'}|$ **do**
3:    $i^{c'} \leftarrow I^{c'}[k]$
4:    $o_1 \leftarrow$ any feature instance in $i^{c'}$
5:    $N \leftarrow T_{o_1}^{\hat{c}}$
6:    **for** $o \in i^{c'} \setminus \{o_1\}$ **do**
7:      $N \leftarrow N \cap T_o^{\hat{c}}$
8:      **if** $N = \emptyset$ **then break**
9:    **for** $o^{\hat{c}} \in N$ **do**
10:     $i^c \leftarrow i^{c'} \cup \{o^{\hat{c}}\}$
11:     append i^c to the list I^c

---

in $i^{c'}$ without having to access all the limited neighborhoods. Thus, this allows to reduce the complexity of the candidate instance identification process.

Improved candidate instance identification algorithm is presented in Algorithm 2. The algorithm requires mostly the same input as Algorithm 1. However, it maintains additional lists for storing CBBs of candidate instances. As an input the list $B^{c'}$ of CBBs of instances of candidate's prefix are needed. As a result, along with the list I^c, a list B^c of CBBs of candidate's instances is computed.

First, lists I^c and B^c are initialized (line 1). In the main loop, (lines 2–11) we iterate over all instances of the candidate's prefix. The prefix instance $i^{c'}$ is retrieved in line 3. Next, the corresponding CBB is retrieved in line 4. Now, to produce all the candidate's instances, we must obtain a list of potential common neighbors with feature c'. For this purpose, a limited neighborhood $T_{o_1}^{\hat{c}}$ of any $o_1 \in i^{c'}$ is sufficient. The random object o_1 is chosen in line 5. The objects $o^{\hat{c}} \in T_{o_1}^{\hat{c}}$ are verified whether they are common neighbors of objects in $i^{c'}$ in the inner loop (lines 6–11). First, the CBB $M(i^c)$ of the potential candidate instance is computed (line 9). If it is not empty, then all objects in $i^c = i^{c'} \cup \{o^{\hat{c}}\}$ are neighbors and thus i^c is a co-location candidate instance. In such a case, the results are appended to the lists I^c and B^c (lines 10–11).

Let us now determine the complexity of the improved algorithm. Two operations are performed in the inner loop: finding the intersection of the $M(i^{c'})$ CBB and $M(o^{\hat{c}})$ OBB, and storing the results. While storing the results can be treated as constant time, finding intersection of two bounding boxes involves intersecting intervals for every dimension and thus requires m steps. Hence, the complexity of the inner loop is $O(m|T_{o_1}^{\hat{c}}|)$. The outer loop is repeated $|I^{c'}|$ times and thus the complexity of the new algorithm is $O(|I^{c'}|m|T_{o_1}^{\hat{c}}|)$.

Algorithm 2. Improved instance identification

input:
    a co-location candidate c
    a list of instances of the candidate's prefix $I^{c'}$ and corresponding list of CBBs $B^{c'}$
    an iCPI-tree for quick retrieval of T_o^f

output:
    a list of the candidate's instances I^c and corresponding list of CBBs B^c

1: $I^c \leftarrow \emptyset,\ B^c \leftarrow \emptyset$
2: **for** $k \in 1, 2, \ldots, |I^{c'}|$ **do**
3:      $i^{c'} \leftarrow I^{c'}[k]$
4:      $M(i^{c'}) \leftarrow B^{c'}[k]$
5:      $o_1 \leftarrow$ any feature instance in $i^{c'}$
6:      **for** $o^{\hat{c}} \in T_{o_1}^{\hat{c}}$ **do**
7:          $M(i^c) \leftarrow M(i^{c'}) \cap M(o^{\hat{c}})$
8:          **if** $M(i^c)$ is not empty **then**
9:              $i^c \leftarrow i^{c'} \cup \{o^{\hat{c}}\}$
10:             append i^c to the list I^c
11:             append $M(i^c)$ to the list B^c

The original algorithm's complexity is $O(|I^{c'}||i^{c'}||T_o^{\hat{c}}|)$. Hence, the two complexities differ only by the second term: $|i^{c'}|$ vs m. It is easy to notice that only in sporadic cases m is greater than 3 and most typically, it is equal to 2. Moreover, it is constant throughout the whole mining process. On the other hand, $|i^{c'}|$ is equal to $1, 2, 3\ldots$ in consecutive iterations of the main apriori loop. Hence, unless the co-locations in the dataset are very small (and a small number iterations is required), using the improved algorithm will provide higher performance.

The new approach also allows for reducing the number of iterations of the inner loop (Algorithm 2, line 6). Let us assume that limited neighborhoods are sorted by coordinates of one dimension (say dimension j). Instead of testing all potential neighbors, we can binary search for the first neighbor $o^{\hat{c}}$ such, that upper bound in $M(o^{\hat{c}})$ on the j dimension is greater or equal to the lower bound in $M(i^{c'})$ (i.e. the first potentially intersecting bounding box). Moreover, the loop can be terminated once the upper bound in $M(i^{c'})$ is greater than the lower bound in $M(o^{\hat{c}})$, since no subsequent $o^{\hat{c}}$ can be a neighbor. This optimization does not change the algorithm's complexity (since in the worst case the whole limited neighborhood must be checked regardless), but it still improves the overall performance.

5   Experiments

To verify the performance of the proposed solution (with the binary search optimization), we have compared it with the state-of-the-art iCPI-tree approach. In the presented figures, we call these approaches BB-identification and iCPI-identification respectively. Experiments have been conducted on a PC (Intel Core I7-6700 3.4 GHz CPU, 32 GB RAM) using C++ implementation of the algorithms. We examined processing times using two real-world datasets. The first dataset, POL, is based on the data from Police Department Incident Reports for San Francisco[1]. It consists of 19594 records describing 35 types (i.e. 35 spatial features) of criminal incidents from the year 2017. The second dataset, OSM, is based on the data from OpenStreetMap project (https://planet.openstreetmap.org/). OSM dataset was prepared by: (1) choosing 100 features corresponding to node types in OpenStreetMap data, (2) randomly choosing a position *op* in USA territory, (3) selecting 180K nearest objects with chosen features around position *op*. Comparing these two datasets, we can state that POL is a small

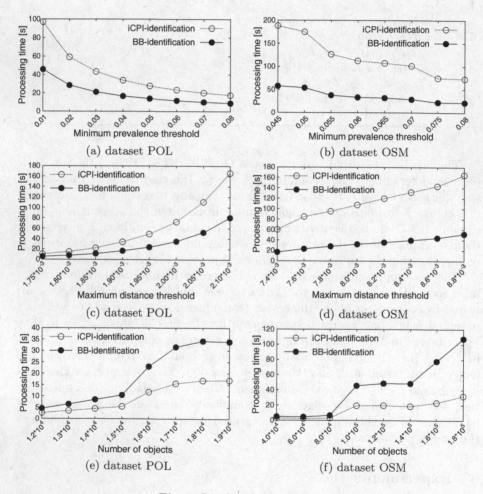

Fig. 4. Results of experiments

[1] https://data.sfgov.org/Public-Safety/Police-Department-Incident-Reports-Historical-2003/tmnf-yvry.

dataset with a low number of features while OSM consists of a much higher number of objects and features, however placed on a larger area.

In the first experiment, we examined how varying the minimum prevalence threshold ($minprev$) affects the performance of both solutions. The results are presented in Fig. 4a and in Fig. 4b for POL and OSM datasets respectively. In both cases, we used constant values of maximum distance thresholds (d). The values for both parameters, $minprev$ and d, have been experimentally selected in such a way that we can illustrate the most interesting changes in the performance of examined algorithms. In the first experiment, d was set to 0.0019 for POL and 0.008 for OSM. For both datasets, the BB-identification method is much faster than the iCPI-identification method. As expected, the higher $minprev$, the lower the processing times. However, it is important to notice that the performance gap strongly increases with decreasing $minprev$. Lower $minprev$ results not only in a higher number of co-locations, but it also impacts the length of the candidates during the mining process. It is clearly observable that for the lowest values of $minprev$ used in experiments, the speedup over state-of-the-art approach is the biggest. The BB-identification method was up to 3.4 times faster than the iCPI-identification algorithm.

In the second experiment, $minprev$ was set to the constant value (0.04 for POL and 0.065 for OSM), while d values were varied in a specified range. The results are presented in Fig. 4c (POL) and in Fig. 4d (OSM). Once again, the BB-identification method is faster than the iCPI-identification. This time however, the speedups are more constant across the whole experiment. For OSM, increasing d results even in a slight decrease of speedups (from 3.6 times faster for low values of d to 3.2 times faster for high values of d). In general, increasing d results in a higher number of neighborhood relationships, thus more instances of candidates can be discovered. Recall that the proposed solution should be the most effective in identification instances for long candidates. Increasing d indirectly affects prevalences (more objects taking part in candidate's instances) and size of generated candidates, however not so strongly as direct manipulation of $minprev$.

In the last experiment, we examined the influence of the number of objects on the performance of both methods. For each dataset, we calculated the middle point and then we removed the appropriate number of objects with the highest distances from that middle point. For dataset POL (Fig. 4e), we limited the number of objects to 12K (step 1K), while for dataset OSM (Fig. 4f) it was 40K (step 20K). It is clearly visible that the number of objects has a major influence on the performance of both algorithms. For a lower number of objects, processing times are very small, however the BB-identification method is still faster than the iCPI-identification method. We can also notice, that for real-world data, there are some spots on the map that can strongly affect the performance. Therefore, removing exactly the same number of objects can impact the processing time in different ways, which is visible in Fig. 4e as well as in Fig. 4f.

6   Summary

In this paper, we proposed a new representation of co-location instances based on bounding boxes and we formulated theorems regarding such a representation that can increase the performance of CPM. We utilized those concepts in the novel algorithm that we compared with the state-of-the-art iCPI-based method. The computational complexity and gathered results from experiments on two real-world datasets show that the proposed solution offers higher efficiency than the currently available method.

Acknowledgement. This research has been partially supported by the statutory funds of Poznan University of Technology.

References

1. Shekhar, S., Huang, Y.: Discovering spatial co-location patterns: a summary of results. In: Jensen, C.S., Schneider, M., Seeger, B., Tsotras, V.J. (eds.) SSTD 2001. LNCS, vol. 2121, pp. 236–256. Springer, Heidelberg (2001). https://doi.org/10.1007/3-540-47724-1_13
2. Prasanthi, G., Maha Lakshmi, J., Raj Kumar, M., Sridhar Babu, N.: Identification of epidemic dengue through spatial mining and visualization with map server. Int. J. Eng. Res. Technol. **1**(7) (2012). ISSN 2278–0181
3. Zhou, G., Zhang, R., Shi, Y., Su, C., Liu, Y., Yan, H.: Extraction of exposed carbonatite in karst desertification area using co-location decision tree. In: Proceedings of the International Geoscience and Remote Sensing Symposium 2014 (IGARSS 2014), July 2014, pp. 3514–3517 (2014)
4. Li, J., Zaïane, O.R., Osornio-Vargas, A.: Discovering statistically significant co-location rules in datasets with extended spatial objects. In: Bellatreche, L., Mohania, M.K. (eds.) DaWaK 2014. LNCS, vol. 8646, pp. 124–135. Springer, Cham (2014). https://doi.org/10.1007/978-3-319-10160-6_12
5. Agrawal, R., Srikant, R.: Fast algorithms for mining association rules in large databases. In: Proceedings of the 20th International Conference on Very Large Data Bases (VLDB 1994), San Francisco, pp. 487–499. Morgan Kaufmann Publishers Inc (1994)
6. Yoo, J.S., Shekhar, S.: A partial join approach for mining co-location patterns. In: Pfoser, D., Cruz, I.F., Ronthaler, M. (eds.) Proceedings of the 12th Annual ACM International Workshop on Geographic Information Systems (GIS 2004), pp. 241–249. ACM, New York (2004)
7. Yoo, J.S., Shekhar, S.: A joinless approach for mining spatial colocation patterns. IEEE Trans. Knowl. Data Eng. **18**(10), 1323–1337 (2006)
8. Wang, L., Bao, Y., Joan, L.: Efficient discovery of spatial co-location patterns using the iCPI-tree. Open Inf. Syst. J. **3**(2), 69–80 (2009)
9. Andrzejewski, W., Boinski, P.: Efficient spatial co-location pattern mining on multiple GPUs. Expert Syst. Appl. **93**(Supplement C), 465–483 (2018)
10. Freeman, H., Shapira, R.: Determining the minimum-area encasing rectangle for an arbitrary closed curve. Commun. ACM **18**(7), 409–413 (1975)
11. Manolopoulos, Y., Nanopoulos, A., Papadopoulos, A.N., Theodoridis, Y.: R-Trees: Theory and Applications. Advanced Information and Knowledge Processing, Springer, London (2010)

Benchmarking Data Lakes Featuring Structured and Unstructured Data with DLBench

Pegdwendé N. Sawadogo[✉] and Jérôme Darmont

Université de Lyon, Lyon 2, UR ERIC, 5 Avenue Pierre Mendès France,
F69676 Bron Cedex, France
{pegdwende.sawadogo,jerome.darmont}@univ-lyon2.fr

Abstract. In the last few years, the concept of data lake has become trendy for data storage and analysis. Thus, several approaches have been proposed to build data lake systems. However, these proposals are difficult to evaluate as there are no commonly shared criteria for comparing data lake systems. Thus, we introduce DLBench, a benchmark to evaluate and compare data lake implementations that support textual and/or tabular contents. More concretely, we propose a data model made of both textual and CSV documents, a workload model composed of a set of various tasks, as well as a set of performance-based metrics, all relevant to the context of data lakes. As a proof of concept, we use DLBench to evaluate an open source data lake system we previously developed.

Keywords: Data lakes · Benchmarking · Textual documents · Tabular data

1 Introduction

Over the last decade, the concept of data lake has emerged as a reference for data storage and exploitation. A data lake is a large repository for storing and analyzing data of any type and size, kept in their raw format [3]. Data access and analyses from data lakes largely rely on metadata [12], making data lakes flexible enough to support a broader range of analyses than traditional data warehouses. Data lakes are thus handy for both data retrieval and data content analysis.

However, the concept of data lake still lacks standards [15]. Thus, there is no commonly shared approach to build, nor to evaluate a data lake. Moreover, existing data lake architectures are often evaluated in diverse and specific ways, and are hardly comparable with each other. Therefore, there is a need of benchmarks to allow objective and comparative evaluation of data lake implementations. There are several benchmarks for big data systems in the literature, but none of them considers the wide range of possible analyses in data lakes.

© Springer Nature Switzerland AG 2021
M. Golfarelli et al. (Eds.): DaWaK 2021, LNCS 12925, pp. 15–26, 2021.
https://doi.org/10.1007/978-3-030-86534-4_2

Hence, we propose in this paper the Data Lake Benchmark (DLBENCH) to evaluate data management performance in data lake systems. We particularly focus in this first instance on textual and tabular contents, which are often included in data lakes. DLBENCH is data-centric, i.e., it focuses on a data management objective, regardless of the underlying technologies [2]. We also designed it with Gray's criteria for a "good" benchmark in mind, namely relevance, portability, simplicity and scalability [8].

More concretely, DLBENCH features a data model that generates textual and tabular documents. By tabular documents, we mean spreadsheet or Comma Separated Value (CSV) files whose integration and querying is a common issue in data lakes. A scale factor parameter SF allows to vary data size in predetermined proportions. DLBENCH also features a workload model, i.e., a set of analytical operations relevant to the context of data lakes with textual and/or tabular content. Finally, we propose a set of performance-based metrics to evaluate such data lake implementations, as well as an execution protocol to execute the workload model on the data model and compute the metrics.

The remainder of this paper is organized as follows. In Sect. 2, we show how DLBENCH differs from existing benchmarks. In Sect. 3, we provide DLBENCH's full specifications. In Sect. 4, we exemplify how DLBENCH works and the insights it provides. Finally, in Sect. 5, we conclude this paper and present research perspectives.

2   Related Works

Benchmarking data lakes mainly relates to two benchmark categories, namely big data and text benchmarks. In this section, we present recent works in these categories and discuss their limitations with respect to our benchmarking objectives.

2.1   Big Data Benchmarks

Big data systems are so diverse that each of big data benchmarks in the literature only target a part of big data requirements [1]. The *de facto* standard TPC-H [18] and TPC-DS [20] issued by the Transaction Processing Performance Council are still widely used to benchmark traditional business intelligence systems. They provide data models that reflect a typical data warehouse, as well as a set of typical business queries, mostly in SQL. BIGBENCH [7] is another reference benchmark that addresses SQL querying on data warehouses. In addition, BIGBENCH adaptations [6,10] include more complex big data analysis tasks, namely sentiment analysis over short texts.

TPCx-HS [19] is a quite different benchmark that aims to evaluate systems running on Apache Hadoop[1] or Spark[2]. For this purpose, only a sorting workload helps measuring performances. HIBENCH [9] also evaluates Hadoop/Spark

---

[1] https://hadoop.apache.org/.
[2] http://spark.apache.org/.

systems, but with a broader range of workloads, i.e., ten workloads including SQL aggregations and joins, classification, clustering and sorts [11]. In the same line, TPCx-AI [21], which is still in development, includes more analysis tasks relevant to big data systems, such as advanced machine learning tasks for fraud detection and product rating.

2.2 Textual Benchmarks

In this category, we consider big data benchmarks with a consequent part of text on one hand, and purely textual benchmarks on the other hand. BIG-DATABENCH [23] is a good representative of the first category. It indeed includes a textual dataset made of Wikipedia[3] pages, as well as classical information retrieval workloads such as *Sort, Grep* and *Wordcount* operations.

One of the latest purely textual benchmarks is TEXTBENDS [22], which aims to evaluate performances of text analysis and processing systems. For this purpose, TEXTBENDS proposes a tweet-based data model and two types of workloads, namely *Top-K keywords* and *Top-K documents* operations. Other purely textual benchmarks focus on language analysis tasks, e.g., Chinese [25] and Portuguese [5] text recognition, respectively.

2.3 Discussion

None of the aforementioned benchmarks proposes a workload sufficiently extensive to reflect all relevant operations in data lakes. In the case of structured data, most benchmark workloads only consider SQL operations (TPC-H, TPC-DS, HIBENCH). More sophisticated machine learning operations remain marginal, while they are common analyses in data lakes. Moreover, the task of finding related data (e.g., joinable tables) is purely missing, while it is a key feature of data lakes.

Existing textual workloads are also insufficient. Admittedly, BIG-DATABENCH's *Grep* and TEXTBENDS's *Top-K documents* operation are relevant for data search. Similarly, *Top-K keywords* and *WordCount* are relevant to assess documents aggregation [9, 22]. However, other operations such as finding most similar documents or clustering documents should also be considered.

Thus, our DLBENCH benchmark stands out, with a broader workload that features both data retrieval and data content analysis operations. DLBENCH's data model also differs from most big data benchmarks as it provides raw tabular files, inducing an additional data integration challenge. Moreover, DLBENCH includes a set of long textual documents that induces a different challenge than short texts such as tweets [22] and Wikipedia articles [23]. Finally, DLBENCH is data-centric, unlike big data benchmarks that focus on a particular technology, e.g., TPCx-HS and HIBENCH.

---

[3] https://en.wikipedia.org/.

3   Specification of DLBench

In this section, we first provide a thorough description of DLBench's data and workload model. Then, we propose a set of metrics and introduce an assessment protocol to evaluate and/or compare systems using DLBench.

3.1   Data Model

Data Description. DLBench includes two types of data to simulate a data lake: textual documents and tabular data. Textual documents are scientific articles that span from few to tens of pages. They are written in French and English and their number amounts to 50,000. Their overall volume is about 62 GB.

Tabular data are synthetically derived from a few CSV files containing Canadian government open data. Although such data are often considered as structured, they still need integration to be queried and analyzed effectively. DLBench features up to 5,000 tabular files amounting to about 1,4 GB of data.

The amount of data in the benchmark can be customised through scale factor parameter SF, which is particularly useful to measure a system's performance when data volume increases. SF ranges from 1 to 5. Table 1 describes the actual amount of data obtained with values of SF.

Table 1. Amount of data per SF value

Scale factors	$SF = 1$	$SF = 2$	$SF = 3$	$SF = 4$	$SF = 5$
Nb. of textual documents	10,000	20,000	30,000	40,000	50,000
Nb. of tabular files	1,000	2,000	3,000	4,000	5,000
Textual documents' size (GB)	8.0	24.9	37.2	49.6	62.7
Tabular files' size (GB)	0.3	0.6	0.8	1.1	1.4

DLBench's data come with metadata catalogues that can serve in data integration. More concretely, we generate from textual documents catalogue information on *year*, *language* and *domain* (discipline) to which each document belongs. Similarly, we associate in the tabular file catalogue a *year* with each file. This way, we can separately query each type of data through its specific metadata. We can also jointly query textual documents and tabular files through the *year* field.

Eventually, textual documents are generated independently from tabular files. Therefore, each type of data can be used apart from the other. In other words, DLBench can be used to assess a system that contains either textual documents only, tabular files only, or both. When not using both types of data, the workload model must be limited to its relevant part.

Data Extraction. We extract textual data from HAL[4], a French open data repository dedicated to scientific document diffusion. We opted for scientific documents as most are long enough to provide complexity and reflect most actual use cases in textual data integration systems, in contrast with shorter documents such as reviews and tweets.

Although HAL's access is open, we are not allowed to redistribute data extracted from HAL. Thus, we provide instead a script that extracts a user-defined amount of documents. This script and a usage guide are available online for reuse[5]. Amongst all available documents in HAL, we restrict to scientific articles whose length is homogeneous, which amounts to 50,000 documents. While extracting documents, the script also generates the metadata catalogue described above.

Tabular data are reused from an existing benchmark [13]. These are actually a set of 5,000 synthetic tabular data files generated from an open dataset stored inside a SQLite[6] database. Many of the columns in the tables contain similar data and can therefore be linked, making this dataset suitable to assess structured data integration as performed in data lakes.

We apply on this original dataset[7] a script to extract all (or a part) of the tables in the form of raw CSV files. As for textual documents, this second script also generates a metadata catalogue. The script as well as guidelines are available online (see footnote 5).

3.2 Workload Model

To assess and compare data lakes across different implementations and systems, some relevant tasks are needed. Thus we specify in this section instances of probable tasks in textual and tabular data integration systems. Furthermore, we translate each task into concrete, executable queries (Table 2).

Data Retrieval Tasks are operations that find data bearing given characteristics. Three main ways are usually exploited to retrieve data in a lake. They are relevant for both tabular data and textual documents. However, we mainly focus on data retrieval from textual documents, as they represent the largest amount of data.

1. *Category filters* consist in filtering data using tags or data properties from the metadata catalogue. In other words, it can be viewed as a navigation task.
2. *Term-based search* pertains to find data, with the help of an index, from all data files that contain a set of keywords. Keyword search is especially relevant for textual documents, but may also serve to retrieve tabular data.

---

[4] https://hal.archives-ouvertes.fr/.
[5] https://github.com/Pegdwende44/DLBench.
[6] https://www.sqlite.org/.
[7] https://storage.googleapis.com/table-union-benchmark/large/benchmark.sqlite.

3. *Related data search* aims to, from a specified data file, retrieve similar data. It can be based on, e.g., column similarities in tabular data, or semantic similarity between textual documents.

Textual Document Analysis/Aggregation tasks work on data contents. Although textual documents and tabular data can be explored with the same methods, they require more specific techniques to be jointly analyzed or aggregated in data lakes.

4. *Document scoring* is a classical information retrieval task that consists in providing a score for each document with respect to how it matches a set of terms. Such scores can be calculated by diverse ways, e.g., with the Elastic-Search [4] scoring algorithm. In all cases, scores depend on the appearance frequency of query terms in the document to score and in the corpus, and also the document's length. This operation is actually very similar to computing top-k documents.

5. *Document highlights* extract a concordance from a corpus. A concordance is a list of snippets where a set of terms appear. It is also a classical information retrieval task that provides a sort of summary of documents.

6. *Document top keywords* are another classical way to summarize and aggregate documents [14]. Computing top keywords is thus a suitable task to assess systems handling textual documents.

7. *Document text mining.* In most data lake systems, data are organized in collections, using tags for example. Here, we propose a data mining task that consists either in representing each collection of documents with respect to the others, or in grouping together similar collections with respect to their intrinsic vocabularies. In the first case, we propose a Principal Component Analysis (PCA) [24] where statistical individuals are document collections. PCA could, for example, out put an average bag of words for each collection. In the second case, we propose a KMeans [17] clustering to detect groups of similar collections.

Tabular Data Analysis/Queries. Finally, we propose specific tasks suitable for integrated tabular files.

8. *Simple Table Queries.* We first propose to evaluate a data lake system's capacity to answer simple table queries through a query language such as SQL. As we are in a context of raw tabular data, language-based querying is indeed an important challenge to address.

9. *Complex Table Queries.* In line with the previous task, we propose to measure how the system supports advanced queries, namely join and grouping queries.

10. *Tuple Mining.* An interesting way to analyze tabular data is either to represent each row with respect to the others or to group together very similar rows. We essentially propose here the same operation as Task #7 above, except that statistical individuals are table rows instead of textual documents. To achieve such an analysis, we only consider numeric values.

Table 2. Query instances

Task	Query	
Data retrieval		
#1	Q1a	Retrieve documents written in *French*
	Q1b	Retrieve documents written in *English* and edited in *December*
	Q1c	Retrieve documents whose domains are *math* or *info*, written in *English* and edited in *2010, 2012* or *2014*
#2	Q2a	Retrieve data files (documents or tables) containing the term *university*
	Q2b	Retrieve data files containing the terms *university, science* or *research*
#3	Q3a	Retrieve the top 5 documents similar to any given document
	Q3b	Retrieve 5 tables joinable to table $t\_dc9442ed0b52d69c\_\_\_\_c11\_1\_\_\_\_1$
Textual Document Analysis/Aggregation		
#4	Q4a	Calculate documents scores w.r.t. the terms *university* and *science*
	Q4b	Calculate documents scores w.r.t. the terms *university, research, new* and *solution*
#5	Q5a	Retrieve documents concordance w.r.t. the terms *university* and *science*
	Q5b	Retrieve documents concordance w.r.t. the terms *university, science new* and *solution*
#6	Q6a	Find top 10 keywords from all documents (stopwords excluded)
#7	Q7a	Run a PCA with documents merged by *domains*
	Q7b	Run a 3-cluster KMeans clustering with documents merged by *domains*
Tabular data analysis/queries		
#8	Q8a	Retrieve all tuples from table $t\_e9efd5cda78af711\_\_\_\_c11\_1\_\_\_\_1$
	Q8b	Retrieve tuples from table $t\_e9efd5cda78af711\_\_\_\_c11\_1\_\_\_\_1$ whose column *PROVINCE* bears the value *BC*
#9	Q9a	Calculate the average of columns *Unnamed: 12, 13,* and *20* from table $t\_356fc1eaad97f93b\_\_\_\_c15\_1\_\_\_\_1$ grouped by *Unnamed: 2*
	Q9b	Run a left join query between tables $PED\_SK\_DTL\_SNF\_\_\_\_c7\_0\_\_\_\_1$ and $t\_285b3bcd52ec0c86\_\_\_\_c13\_1\_\_\_\_1$ w.r.t. columns named *SOILTYPE*
#10	Q10a	Run a PCA on the result of query *Q9a*
	Q10b	Run a 3-cluster KMeans clustering on the result of query *Q9a*

3.3   Performance Metrics

In this section, we propose a set of three metrics to compare and assess data lake implementations.

1. **Query execution time** aims to measure the time necessary to run each of the 20 query instances from Table 2 on the tested data lake architecture. This metric actually reports how efficient the lake's metadata system is, as it serves to integrate raw data, and thus make analyses easier and faster. In the case where certain queries are not supported, measures are only computed on the supported tasks.
2. **Metadata size** measures the amount of metadata generated by the system. It allows to balance the execution time with the resulting storage cost.

3. **Metadata generation time** encompasses the generation of all the lake's metadata. This also serves to balance query execution time.

   We did not include other possible metrics such as actually used RAM and CPU because they are hard to measure. However, we recommend interpreting benchmark results while taking into account available RAM and CPU.

3.4   Assessment Protocol

The three metrics from Sect. 3.3 are measured through an iterative process for each scale factor $SF \in \{1, 2, 3, 4, 5\}$. Each iteration consists in four steps (Algorithm 1).

1. **Data generation** is achieved with the scripts specified in Sect. 3.1 (see footnote 5).
2. **Data integration.** Raw data now need to be integrated in the data lake system through the generation and organization of metadata. This step is specific to each system, as there are plethora of ways to integrate data in a lake.
3. **Metadata size and generation time computing** consists in measuring metrics the total size of generated metadata and the time taken to generate all metadata, with respect to the current SF.
4. **Query execution time computation** involves computing the running time of each individual query. To mitigate any perturbation, we average the time of 10 runs for each query instance. Let us notice that all timed executions must be warm runs, i.e., each of the 20 query instances must first be executed once (a cold run not taken into account in the results).

Algorithm 1: Assessment protocol

Result: metric_1, metric_2, metric_3
metric_1 ← [ ][ ]; metric_2 ← [ ]; metric_3 ← [ ];
for *SF ← 1 to 5* **do**
    generate_benchmark_data(SF);
    generate_and_organize_metadata(SF);
    metric_2[SF] ← retrieve_metadata_generation_time(SF);
    metric_3[SF] ← retrieve_metadata_size(SF);
    for *i ← 1 to 20* **do**
        run_query(i, SF);
        response_times ← [ ];
        for *j ← 1 to 10* **do**
            response_times[j] ← run_query(i, SF);
        end
        metric_1[SF][i] ← average(response_times);
    end
end

---

4   Proof of Concept

4.1   Overview of AUDAL

To demonstrate the use of DLBENCH, we evaluate AUDAL [16], a data lake system designed as part of a management science project, to allow automatic and advanced analyses on various textual documents (annual reports, press releases, websites, social media posts...) and spreadsheet files (information about companies, stock market quotations...). The AUDAL system uses an extensive metadata system stored inside MongoDB[8], Neo4J[9], SQLite (see footnote 6) and ElasticSearch[10] to support numerous analyses.

AUDAL provides ready-to-use analyses via a representational state transfer application programming interface (REST API) dedicated to data scientists, and also through a Web-based analysis platform designed for business users.

4.2   Setup and Results

AUDAL is implemented on a cluster of three VMware virtual machines (VMs). The first VM has a 7-core Intel-Xeon 2.20 GHz processor and 24 GB of RAM. It runs the API and also supports metadata extraction. Both other VMs have a mono-core Intel-Xeon 2.20 GHz processor and 24 GB of RAM. Each of the three VMs hosts a Neo4J instance, an ElasticSearch instance and a MongoDB instance to store AUDAL's metadata.

The results achieved with DLBENCH show that AUDAL scales quite well (Figs. 1, 2, 4, 5 and 6). Almost all task response times are indeed either constant (Tasks #3, #7, #8, #9 and #10) or grow linearly with SF (Tasks #1, #2, and #4 to #6). In addition, we observe that except Task #6 (that takes up to 92 s), all execution times are reasonable considering the modest capabilities of our hardware setup. Eventually, metadata generation time and size scale linearly and almost-linearly, respectively (Fig. 7). We can also see that metadata amount to about half the volume of raw data, which illustrates how extensive AUDAL's metadata are. We observe some fluctuations in the results, with sometimes negative slopes while SF increases. Such variations are due to external, random factors such as network load or the Java garbage collector starting running. However, the influence on the runtime is negligible (of the order of a tenth of a second) and is only visible on simple queries that run in half a second.

---

[8] https://www.mongodb.com.
[9] https://neo4j.com.
[10] https://www.elastic.co.

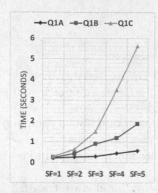

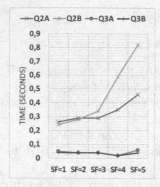

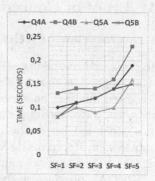

Fig. 1. Task #1 response times

Fig. 2. Tasks #2 & #3 response times

Fig. 3. Tasks #4 & #5 response times

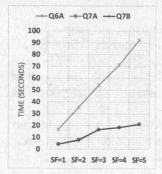

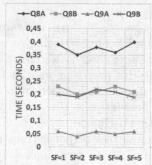

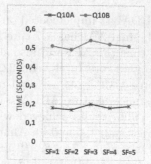

Fig. 4. Tasks #6 & #7 response times

Fig. 5. Tasks #8 & #9 response times

Fig. 6. Task #10 response times

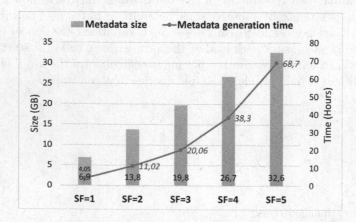

Fig. 7. Metadata size and generation time

5   Conclusion

In this paper, we introduce DLBench, a benchmark for data lakes with textual and/or tabular contents. To the best of our knowledge, DLBench is the first data lake benchmark. DLBench features: 1) a data model made of a corpus of long, textual documents on one hand, and a set of raw tabular data on the other hand; 2) a query model of twenty query instances across ten different tasks; 3) three relevant metrics to assess and compare data lake implementations; and 4) an execution protocol. Finally, we demonstrate the use of DLBench by assessing the AUDAL data lake [16], highlighting that the AUDAL system scales quite well, especially for data retrieval and tabular data querying.

Future works include an extension of the structured part of DLBench's data model with an alternative, larger dataset. Another enhancement of DLBench could consists in providing an overview of value distributions in generated data. Finally, we plan to perform a comparative study of existing data lake systems using DLBench.

Acknowledgments. P.N. Sawadogo's PhD is funded by the Auvergne-Rhöne-Alpes Region through the AURA-PMI project.

References

1. Bajaber, F., Sakr, S., Batarfi, O., Altalhi, A.H., Barnawi, A.: Benchmarking big data systems: l survey. Comput. Commun. **149**, 241–251 (2020). https://doi.org/10.1016/j.comcom.2019.10.002
2. Darmont, J.: Data-centric benchmarking. In: Advanced Methodologies and Technologies in Network Architecture, Mobile Computing, and Data Analytics, pp. 342–353. IGI Global, Hershey (2019)
3. Dixon, J.: Pentaho, Hadoop, and Data Lakes (2010). https://jamesdixon.wordpress.com/2010/10/14/pentaho-hadoop-and-data-lakes/
4. Elasticsearch: Theory Behind Relevance Scoring (2019). https://www.elastic.co/guide/en/elasticsearch/guide/current/scoring-theory.html
5. Fialho, P., Coheur, L., Quaresma, P.: Benchmarking natural language inference and semantic textual similarity for portuguese. Information **11**(10), 484 (2020). https://doi.org/10.3390/info11100484
6. Ghazal, A., et al.: Bigbench V2: the new and improved bigbench. In: 33rd IEEE International Conference on Data Engineering (ICDE 2017), San Diego, CA, USA, April 19–22 2017. pp. 1225–1236. IEEE Computer Society (2017). https://doi.org/10.1109/ICDE.2017.167
7. Ghazal, A., et al.: Bigbench: towards an industry standard benchmark for big data analytics. In: Ross, K.A., Srivastava, D., Papadias, D. (eds.) Proceedings of the ACM SIGMOD International Conference on Management of Data (SIGMOD 2013), New York, June 22–27, 2013. pp. 1197–1208. ACM (2013). https://doi.org/10.1145/2463676.2463712
8. Gray, J.: Database and Transaction Processing Performance Handbook (1993). http://jimgray.azurewebsites.net/benchmarkhandbook/chapter1.pdf

9. Huang, S., Huang, J., Dai, J., Xie, T., Huang, B.: The hibench benchmark suite: characterization of the mapreduce-based data analysis. In: 2010 IEEE 26th International Conference on Data Engineering Workshops (ICDEW 2010), pp. 41–51 (2010). https://doi.org/10.1109/ICDEW.2010.5452747

10. Ivanov, T., Ghazal, A., Crolotte, A., Kostamaa, P., Ghazal, Y.: Corebigbench: Benchmarking big data core operations. In: Tözün, P., Böhm, A. (eds.) Proceedings of the 8th International Workshop on Testing Database Systems (DBTest@SIGMOD 2020), Portland, Oregon, June 19, 2020. pp. 4:1–4:6. ACM (2020). https://doi.org/10.1145/3395032.3395324

11. Ivanov, T., et al.: Big data benchmark compendium. In: Performance Evaluation and Benchmarking: Traditional to Big Data to Internet of Things - 7th TPC Technology Conference (TPCTC 2015), Kohala Coast, HI, USA. pp. 135–155, September 2015. https://doi.org/10.1007/978-3-319-31409-9_9

12. Maccioni, A., Torlone, R.: KAYAK: a framework for just-in-time data preparation in a data lake. In: Krogstie, J., Reijers, H.A. (eds.) CAiSE 2018. LNCS, vol. 10816, pp. 474–489. Springer, Cham (2018). https://doi.org/10.1007/978-3-319-91563-0_29

13. Nargesian, F., Zhu, E., Pu, K.Q., Miller, R.J.: Table union search on open data. In: Proceedings of the VLDB Endowment, vol. 11, pp. 813–825, March 2018. https://doi.org/10.14778/3192965.3192973

14. Ravat, F., Teste, O., Tournier, R., Zurfluh, G.: Top-keyword: an aggregation function for textual document OLAP. In: 10th International Conference on Data Warehousing and Knowledge Discovery (DaWaK 2008), Turin, Italy. pp. 55–64, September 2008. https://doi.org/10.1007/978-3-540-85836-2_6

15. Russom, P.: Data Lakes Purposes. Patterns, and Platforms. TDWI research, Practices (2017)

16. Scholly, E., et al.: Coining goldMEDAL: a new contribution to data lake generic metadata modeling. In: 23rd International Workshop on Design, Optimization, Languages and Analytical Processing of Big Data (DOLAP@EDBT/ICDT 2021), Nicosia, Cyprus. pp. 31–40, March 2021

17. Steinley, D.: K-means clustering: a half-century synthesis. Br. . Math. Stat. Psychol. **59**(1), 1–34 (2006)

18. Transaction Processing Performance Council: TPC Benchmark H - Standard Specification (version 2.18.0) (2014). http://www.tpc.org/tpch/

19. Transaction Processing Performance Council: TPC Express Benchmark HS - Standard Specification (version 2.0.3) (2018). http://www.tpc.org/tpcds/

20. Transaction Processing Performance Council: TPC Benchmark DS - Standard Specification (version 2.13.0) (2020). http://www.tpc.org/tpcds/

21. Transaction Processing Performance Council: TPC Express AI - Draft Specification (version 0.6). http://tpc.org/tpcx-ai/default5.asp (2020)

22. Truică, C.-O., Apostol, E.-S., Darmont, J., Assent, I.: TextBenDS: a generic textual data benchmark for distributed systems. Inf. Syst. Front. **23**(1), 81–100 (2020). https://doi.org/10.1007/s10796-020-09999-y

23. Wang, L., et al.: Bigdatabench: a big data benchmark suite from internet services. In: 2014 IEEE 20th International Symposium on High Performance Computer Architecture (HPCA). pp. 488–499 (2014)

24. Wold, S., Esbensen, K., Geladi, P.: Principal component analysis. Chemom. Intell. Lab. Syst. **2**(1), 37–52 (1987). https://doi.org/10.1016/0169-7439(87)80084-9

25. Zhu, Y., Xie, Z., Jin, L., Chen, X., Huang, Y., Zhang, M.: SCUT-EPT: new dataset and benchmark for offline chinese text recognition in examination paper. IEEE Access **7**, 370–382 (2019). https://doi.org/10.1109/ACCESS.2018.2885398

Towards an Adaptive Multidimensional Partitioning for Accelerating Spark SQL

Soumia Benkrid[1]([⊠]), Ladjel Bellatreche[2], Yacine Mestoui[1], and Carlos Ordonez[3]

[1] Ecole nationale Supérieure d'Informatique (ESI), Oued Smar, Algeria
s_benkrid@esi.dz, fy_mestoui@esi.dz
[2] LIAS/ISAE-ENSMA, Poitiers, France
bellatreche@ensma.fr
[3] University of Houston, Houston, TX, USA
carlos@central.uh.edu

Abstract. Nowadays Parallel DBMSs and Spark SQL compete with each other to query Big Data. Parallel DBMSs feature extensive experience embodied by powerful data partitioning and data allocation algorithms, but they suffer when handling dynamic changes in query workload. On the other hand, Spark SQL has become a solution to process query workloads on big data, outside the DBMS realm. Unfortunately, Spark SQL incurs into significant random disk I/O cost, because there is no correlation detected between Spark jobs and data blocks read from the disk. In consequence, Spark fails at providing high performance in a dynamic analytic environment. To solve such limitation, we propose an adaptive query-aware framework for partitioning big data tables for query processing, based on a genetic optimization problem formulation. Our approach intensively rewrites queries by exploiting different dimension hierarchies that may exist among dimension attributes, skipping irrelevant data to improve I/O performance. We present an experimental validation on a Spark SQL parallel cluster, showing promising results.

Keywords: Multidimensional partitioning · Spark-SQL · Utility maximization · Adaptive query processing · Big data · Dimensional hierarchies

1 Introduction

In the last decades, the digital revolution is evolving at an extreme rate, enabling faster changes and processes, and producing a vast amount of data. In this context, companies are turning urgently to data science to explore, integrate, model, evaluate and automate processes that generate value from data. Beyond actual tools and data details, massive parallel processing is required for providing the scalability that is necessary for any big data application. Indeed, to result in linear speed-up, the DBA may often increase the resource in the cluster by adding

M. Golfarelli et al. (Eds.): DaWaK 2021, LNCS 12925, pp. 27–38, 2021.
https://doi.org/10.1007/978-3-030-86534-4_3

more machines to the computing cluster. Furthermore, this requires additional costs which may be not available in practice.

A promising parallel processing approach requires three main phases: (1) data partitioning and allocating partitions, (2) running processing code on each partition, and (3) gathering or assembling partial results. However, the efficiency of the database parallel processing is significantly sensitive to how the data is partitioned (phase 1). Basically, the partitioning scheme is chosen in a static (offline) environment where the design is done only once and that design can persist. Most commonly, when the change is detected, the DBA must often intervene by taking the entire *DW* offline for repair. The problem is getting bigger and more difficult because the data system has become large and dynamic. Thus, carrying out a redesign at each change is not entirely realistic since the overload of the redesign is likely to be very high. Figure 1 illustrates the time needed to adapt a fragmentation schema among the size of the database. Accordingly, making an adaptive partitioning schema with minimal overload is a crucial issue.

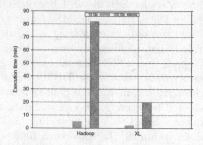

Fig. 1. Fragmentation schema adaptation according the database size

The adaptive issue has been extensively studied by the database community. The first line of works [4,8,10,20,22] focuses on changes in the workload, periodically or when a change occurs, an incremental design is performed. These workload-driven solutions yield promising results, but they require significant time to adapt a fragmentation schema. The second category of works uses Query Adaptive Processing [1,12,14,16,24] that use runtime statistics to choose the most efficient query execution plan during query execution. However, these Data-driven approaches can incur in non-negligible overhead.

As though HDFS stores the data by creating a folder with partition column values, altering the partition schema means the change of the whole directory structure and data. To overcome the problem, Spark SQL a big data processing tool for structured data query and analysis tailored towards using a Data-driven approach that uses runtime statistics to choose the most efficient query execution plan during query execution. Hence, the Spark SQL system is currently leading to serious random disk I/O costs since there is no correlation between the spark jobs and the data to be read from the disk.

In this paper, we introduced a new dynamic SQL framework for Apache Spark. Our novel approach relies on the intuition that multi-dimensional partitioning could accelerate the spark SQL since thereby it supported fewer I/O operations and effective prefetching. Our SQL framework is based on our Genetic Optimization Physical Planner [6] and relies on the intuition that a query should leverage knowledge available in the reference partitioning schema to Spark as much as possible. Specifically, our SQL optimizer (1) uses reference partitioning to enable efficient data processing by avoiding unnecessary read; (2) leverages dimensional hierarchies' information to maximize the benefit of the partitioning schema. The goal of our work is to show that by combining efficient workload-driven approaches and adaptive execution processing, the performance of Spark SQL can be significantly increased.

The remainder of this paper is organized as follows. Related works are discussed in Sect. 2. Section 3 provides the formulation of our studied problem. Section 4 introduces a novel planner for data partitioning based on genetic optimization. Section 5 presents experiments evaluating the quality of our solution. Finally, Sect. 6 summarizes our main findings.

2   Literature Review

Data partitioning has attracted plenty of research attention. Before 2013, researchers typically assume that the query workload is provided upfront [1,7,8, 18,21,25] and try to choose the best partitioning schema in offline (static) mode using mainly some optimization search techniques such as branch-and-bound search [18,25], genetic algorithms [18], and relaxation/approximation [7]. Nevertheless, it is difficult to maintain the usefulness of the offline data partitioning schema in a dynamic context. In fact, there have been two main directions of research, depending on how to select data to be migrated. We can further distinguish between two different categories. 1) Workload driven techniques, which focus on the workload changes and 2) data driven techniques, which work with the optimizer.

Workload-Driven Adaptive Approaches. The most recent works focus on studying how to migrate data based on the workload change [19], and the data items that are affected [4,20,22]. E-STORE [22] is a dynamic partitioning manager for automatically identifying when a reconfiguration is needed using system and database statistics. It explores the idea of managing hot tuples separately from cold tuples by designing a two-tier partitioning method that first distributes hot tuples across the cluster and then allocates cold tuples to fill the remaining space. Simultaneously, the research community has sought to develop additional design techniques based on machine learning, such as view recommendation [11], database cracking [3], workload forecasting [17], cardinality estimation [13] and Autonomous DBMS [2,5,23], by which multiple aspects of self-adaptive can be improved. However, only a few works have recently focused to tune data-partitioning by using reinforcement learning (RL) [8] and deep RL [10]. Hilprecht et al. [10] propose an approach, based on Deep Reinforcement Learning (DRL), in which several DRL agents are trained offline to learn the trade-offs of using different partitioning for a given database schema. For that, they use a cost model to bootstrap the DRL model. If new queries are added to the workload or if the database schema changes, the partitioning agent is adapted by progressive learning. However, to support a completely new database schema, a new set of DRL agents must be trained.

Data-Driven Adaptive Approaches. The data driven adaptive approaches consist of improve the execution performance by generating a new execution plan at runtime if data distributions do not match the optimizer's estimations. Kocsis et al. [14] proposed HYLAS, a tool for optimizing Spark queries using semantic preserving transformation rules to eliminate intermediate superfluous data structures. Zhang et al. [24] propose a system that dynamically generates execution plans at query runtime, and runs those plans on chunks of data. Based on feedback from earlier chunks, alternative plans might be used for later chunks. DynO [12] and Rios [15] focus on adaptive implement by updating the query plan at runtime. Recently, we have proposed a two-step approach based on genetic algorithms to improve the performance of dynamical analytical queries by optimizing data partitioning in a cluster environment[6]. Although this effort put on the query optimizer, it is still in the infant stage for Spark SQL optimizer. This is mainly due to the Parquet storage format, the HDFS block size is much larger, indexes and buffer pool. In this paper, we characterized the effectiveness of the query optimization in the aim of enhancing Spark SQL optimizer with detailed partitioning information. Indeed, our solution aims to filter out most of the records in advance which can reduce the amount of data in the shuffle stage and improve the performance of equivalent connections.

3   Definitions and Problem Formulation

In this section, we present all ingredients that facilitate the formalization of our target problem.

3.1   Preliminaries

Range Referencing Partitioning. In this paper, we reproduce the traditional methodology to partition relational *DW* to HDFS *DW*. More concretely, we *partition some/all dimension tables using the predicates of the workload defined on their attributes, and then partition the fact table based on the partitioning schemes of dimension tables.* To illustrate this partitioning, let us suppose a relational warehouse modelled by a star schema with d dimension tables $(D_1, D_2, ..., D_d)$ and a fact table F. F is the largest table, used on every BI query. Among these dimension tables, g tables are fragmented ($g \leq d$). Each dimension table D_i $(1 \leq i \leq g)$ is partitioned into m_i fragments: $\{D_{i1}, D_{i2}, ..., D_{im_i}\}$, where each fragment D_{ij} is defined as: $D_{ij} = \sigma_{cl_j^i}(D_i)$, where cl_j^i and σ $(1 \leq i \leq g, 1 \leq j \leq m_i)$ represent a conjunction of simple predicates and the selection operator, respectively. Thus, the fragmentation schema of the fact table F is defined as follows: $F_i = F \ltimes D_{1j} \ltimes D_{2k} \ltimes .. \ltimes D_{gl}$, $(1 \leq i \leq m_i)$, where $\ltimes$ represents the semi join operation.

Benefit of a Partitioning Schema for a Query. Using a partitioning schema to answer a query has a significant benefit since it can reduce overall disk performance. Data partitioning decomposes very large tables into smaller partitions, and each partition is an independent object with its own name and its own storage characteristics. The benefit can be calculated by the difference of query cost with/without using the partitioning schema. Which is defined as below.

Definition 1. *Benefit: Given a query q and a partitioning schema SF, the cost of executing q is $cost(q)$, the cost of executing q using SF is $cost(q|SF)$, and the benefit is $B_{q,SF} = cost(q) - cost(q|SF)$. A fragmentation schema should ensure a positive benefit ($B_{q,SF} > 0$).*

Utility of Fragmentation Schema. Given a query q and a partitioning schema SF, we need to compute the utility of using SF for q. The utility of a partitioning schema SF seeks to maximize the benefit of a query, generally, over two periods t and $t + 1$. The utility of a partitioning schema SF over a query q is a metric that measures the reduction threshold in the cost of q by SF. Thus, we define $U(SF, q)$ as follows.

Definition 2. *We suppose that if the period $t + 1$ transmits a new query q, its utility is given by:*

$$U(SF, q) = B_{q,SF'}(q, t+1)/B_{a,SF}(q, t) \tag{1}$$

where $B_{q,SF'}(q, t+1)$ ($\forall i \in \{0, 1\}$) denotes the benefit of q under the partitioning schema SF define at time $t + i$ and SF' is the partitioning schema generated after the consideration of q in the predefined workload.

3.2 Problem Statement

We now describe the formulation of our problem: Given:

- A SPARK-SQL cluster $\mathcal{DBC}$ with M nodes $\mathcal{N} = \{N_1, N_2, \ldots, N_M\}$;
- A relational data warehouse $\mathcal{RDW}$ modeled according to a star schema and composed by one fact table $\mathcal{F}$ and d dimensional tables $\mathcal{D} = \{D_1, D_2, \ldots, D_d\}$.
- a set of star join queries $\mathcal{Q} = \{Q_1, Q_2, \ldots, Q_L\}$ to be executed over $\mathcal{DBC}$, being each query Q_l characterized by an access frequency f_l;
- A fragmentation maintenance constraint W that the designer considers relevant for his/her target partitioning process
- A target profit u which represents the minimum desired utility.

The optimization problem involves finding the best fragmentation schema SF^* such that

$$\text{maximize} \quad \sum_{i=1}^{L} B_{Q_i,SF^\star}$$

$$\text{subject to} \quad |SF^\star| \leq \mathcal{W}, \tag{2}$$

$$\sum_{i \geq 1} \mathcal{U}(SF^\star, Q_i) \geq u.$$

4   System Architecture

Figure 2 illustrates the global architecture of our framework. It contains two main parts. *Partitioning schema selection* and *controller*. First, given a query workload, we determine the best data partitioning schema. Next, the controller interacts with Spark SQL to rewrite queries, launch new configurations, and collects performance measurements. Precisely, the decision-adaptive activity can be regarded as an analysis form that it is trying to choose which adaptation strategy should be started at the current time to maximize the overall utility that the system will provide during the remainder of its execution. Later, we rewrite queries using some highly beneficial of the partitioning schema. As last step, we execute the rewritten workload. Next, we discuss the details these two parts.

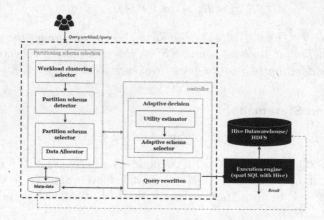

Fig. 2. The overview of system framework.

4.1   Partitioning Schema Selection

The main idea of our partitioning approach, is to build several data partitioning schemes and merge them together to get a more accurate (having high utility) and dynamic global data partitioning schema. Precisely, once the data partitioning schema for each sub-workload is selected, an algorithm is used to aggregate over the data partitioning schemes to form the most efficient data partitioning.

This part contains three components: *workload clustering selector*, *Partitioning Schema detector* and *partitioning schema selector*. At first, we use *workload clustering selector* to divide the given workload to a number of sub-workloads such that queries in the same groups are related to one another. Later, we use *partitioning schema builder* to build a fragmentation schema for each query group. Once the partitioning schemes of the workload clusters have been generated, their outputs must be combined into a single partitioning schema. The *Partitioning Schema selector* merges them together to get a more accurate and stable data partitioning schema. In particular, we use our genetic planner [6]. Finally, the data allocator place the so-generate fragments over Spark-SQL cluster using hash distribution.

4.2 Controller

This part contains two components: *utility estimator* and *adaptive schema selector*. Every query issued by users first goes to the *utility estimator* to compute the utility $U(SF, q)$. For that, we design a cost model to calculate the cost of executing a query on a given fragmentation schema. First, *utility estimator* calls *partitioning schema builder* in order to build un fragmentation schema by considering the new query. After that, we can compute the benefit $B_{q,SF}$ and the utility $U(SF, q)$.

When a violation of performance constraints occurs, the new queries causing the failure are integrated into the fragmentation scheme. Our genetic algorithm is called again to select the best partitioning plan. First, a starting population of adaptation strategies is created based on the last exploration phase population of the genetic algorithm, as well as adaptation strategies for the updated offline classes with ad-hoc queries. Then these adaptation strategies are iteratively improved by applying a combination of mutation and crossover operators, with the most efficient plans being more likely to pass to the next generation. It is expected that this iterative process increases utility over time, thereby reducing average query evaluation. We emphasize that by reusing exploration candidates, we reduce the number of evaluations of the fitness function to choose the best solution.

4.3 Query Rewritten

The *query rewritten* component will extract the dimensions and predicates from the query, and then search the partitioning attribute for any predicate which can provide the data source needed by the query. Upon finding a qualified partitioning attribute, the engine will add/replace the original predicate with the new predicate and then use Spark SQL to execute the query.

In order to reduce the reading of the data, we not only use the fragmentation attributes, but we also use the hierarchy dimension structure [9,21]. We can distinguish the following two main cases for which such an improvement is possible.

Case 1: Queries on "lower-level" attributes of the fragmentation dimension. In fact, each value of an attribute belongs to a low level in the hierarchy corresponds exactly to one value of the fragmentation attribute. If the query references all fragmented dimensions, it requires loading a single fragment to run, otherwise multiple fragments will be loaded. Thus, a rewrite is necessary by adding predicates on the partitioning attributes to load only the valid tuples needed.

Case 2: Queries on "higher-level" attributes of the fragmentation dimension. Queries defined on attributes belonging to a high level in the fragmentation attribute hierarchy also benefit from the fragmentation scheme. This is because the number of fragments to load will be greater than the previous cases, since each value of the attribute has several associated values of the fragmentation attribute. Thus, the number of fragments increases if, and only if, certain dimensions of fragmentation are involved.

Obviously, all queries defined on fragmented dimensions will also benefit from the partitioning schema, as in the case of *Case 1* and *Case 2*. Thus, all queries referencing at least one attribute of a fragmented dimension table benefit from fragmentation by reducing the number of fragments to be processed.

5   Experimental Results

Our experiments were conducted on a Hadoop parallel cluster with 9 nodes, configured as 1 HDFS NameNode (master node) and 8 HDFS DataNodes (workers), the nodes were connected through 100 Mbps Ethernet. All cluster nodes had Linux Ubuntu 16.04.1 LTS as the operating system. On top of that, we installed Hadoop version 3.1.0, which provides HDFS, and Yarn, and Hive 2.0.0, with MySQL database (MySQL 5.5.53) for Hive metadata storage. Separately, we installed Spark 3.1.0 and we configured Spark SQL to work with Yarn and Hive catalog. The overall storage architecture is based on HDFS. First, we load the large amount of data on Hive tables in Parquet columnar format. Then, in the interrogation step, we use Spark SQL to read the Hive partitioned tables We use the well-known Star Schema Benchmark (SSB) to generate the test data set. The size of the data set used is $100\,G$.

In the first experiment, we focused the attention on the impact of fragmentation threshold. The aim of this experiment is to show the importance of choosing the best number of fragment to improve the workload performance. For that, we vary the fragmentation threshold in the interval [100–600] and we calculate the throughput workload execution.

It clearly follows that since the value of W is smaller than 300, increasing of the fragmentation threshold improves the query performance because by releasing the fragmentation threshold, more attributes are used to fragment the warehouse. Although when the value of W is relatively large that decreases the query performance significantly, as HDFS is not appropriate for small data and lacks the ability to efficiently support the random reading of small files because of its high capacity design. This experimental result confirms the importance of choosing the right number of final fragments to generate and online repartitioning may defer further degrade performance.

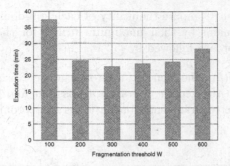

Fig. 3. Effect of the fragmentation threshold W on the query performance.

In the rest of the experiments, we fix the fragmentation threshold (W) to 200. We have chosen to fix W to 200 and not to 300 and this to have a flexibility for the adaptation/repartitionning step.

In the second experiment, we outlined the impact of the using of dimensional hierarchies on query processing. Figure 4 shows the results obtained and confirms that the use of the dimensional hierarchies improves significantly the performance as most of the effective characteristics of queries are taken into consideration. Precisely, our fine-grained rewritten allows many query to be confined too few fragments, thereby reducing I/O. Our Extensive rewritten outperforms Spark SQL optimizer by average 45% in terms of execution time.

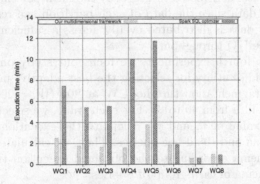

Fig. 4. Impact of using dimensional hierarchies.

To examine the quality of our genetic based approach, we compared our genetic based approach with two others partitioning scheme aggregation approaches by running 100 queries.

– *Majority Voting.* Rank sub-domains by the number of times they appear in different partitioning schema.

- *SUKP-Partionning.* Compute the utility of each sub-domains for each query, and rank sub-domains by their maximum utility for the workload. This is similar to the approach followed by [5].

As shown in Fig.5(a), the genetic approach outperforms the others two approaches. Through the genetic operators (crossover and mutation) the approach gives rise to better fragmentation patterns that promote the majority of queries.

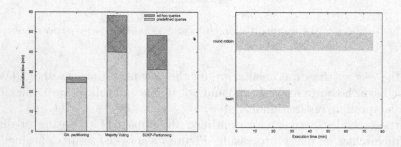

Fig. 5. Impact of the GA on the query performance.

Placing partitions on nodes is essential to achieve good scalability. In order to identify the best allocation mode, we compare the round robbin manner with the hash placement. As depicted in Fig.5(b) the hash placement is an efficient way to distribute data since the data of each partition can reside on all nodes. This ensures that data is distributed evenly across a sufficient number of data nodes to maximize I/O throughput.

Finally, in the fourth experiment, we focus our study on determining the minimum threshold of utility for selecting the best adaptive partitioning. To do this, we set the fragmentation threshold W at 200 ($W = 200$) and we varied the utility threshold u from 40 to 80. For each value, we select a fragmentation schema for each value of u, and we calculate the execution time of the 100 predefined workload and 10 ad-hoc queries. Then, we calculate the utility of the fragmentation schema according to the obtained execution time and the best fragmentation schema of each group of queries.

As depicted in Fig.6, too high or too low utility threshold can significantly impact the performance of the predefined and ad-hoc queries. Precisely, a high threshold may result in not obtaining a high utility fragmentation scheme for ad-hoc queries, but it is sufficiently useful for the predefined workload. Though a low threshold may result in a less meaningful schema for the predefined queries and the schema is perfectly adequate for ad-hoc queries. In addition to this, it should be noted that it is important of carefully choosing the utility desired rate (u). The administrator must carefully choose the best utility threshold according to the frequency of change in the workload.

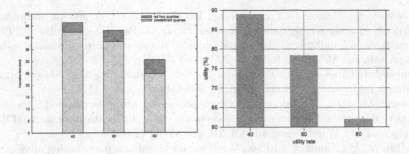

Fig. 6. Impact of the utility threshold on query performance.

6 Conclusions

In this article, we have presented an external adaptive and workload-sensitive partitioning framework for large-scale data. In this framework, we have addressed several performance issues; these include the limitations of static workload-aware partitioning, the overhead of rebuilding partitions in HDFS, and also irrelevant data spark-SQL reads. We have shown that utility-based formalization and the exploitation of partitioning attribute semantics (dimensional hierarchies) successfully address these challenges and provides an efficient Spark-SQL query processing framework. Extensive experiments evaluate partition quality, the impact of parameters, and dimensional hierarchies rewritten efficiency, with encouraging results. We believe that leverages multi-dimensional partitioning with Spark Adaptive Query processing shows promise to analyze dynamic workloads with ad-hoc queries in modern big data environments.

Although the obtained results are interesting and encouraging, this work opens several perspectives: (i) improvement of the adaption policy by using machine learning techniques, (ii) identifying automatically the best Spark SQL configuration (dynamic cluster).

References

1. Akal, F., Böhm, K., Schek, H.-J.: OLAP query evaluation in a database cluster: a performance study on intra-query parallelism. In: ADBIS, pp. 218–231 (2002)
2. Aken, D.V., Pavlo, A., Gordon, G.J., Zhang, B.: Automatic database management system tuning through large-scale machine learning. In: Salihoglu, S., Zhou, W., Chirkova, R., Yang, J., Suciu, D. (eds.) ACM SIGMOD, pp. 1009–1024 (2017)
3. Alagiannis, I., Idreos, S., Ailamaki, A.: H2O: a hands-free adaptive store. In: ACM SIGMOD, pp. 1103–1114 (2014)
4. Asad, O., Kemme, B.: AdaptCache: adaptive data partitioning and migration for distributed object caches. In: Proceedings of the 17th International Middleware Conference, pp. 1–13 (2016)
5. Benkrid, S., Bellatreche, L.: A framework for designing autonomous parallel data warehouses. In: ICA3PP, pp. 97–104 (2019)

6. Benkrid, S., Mestoui, Y., Bellatreche, L., Ordonez, C.: A genetic optimization physical planner for big data warehouses. In: IEEE Big Data, pp. 406–412 (2020)
7. Bruno, N., Chaudhuri, S.: Automatic physical database tuning: a relaxation-based approach. In: ACM SIGMOD, pp. 227–238 (2005)
8. Durand, G.C., et al.: GridFormation: towards self-driven online data partitioning using reinforcement learning. In: aiDM Workshop, pp. 1–7 (2018)
9. Garcia-Alvarado, C., Ordonez, C.: Query processing on cubes mapped from ontologies to dimension hierarchies. In: Proceedings of the Fifteenth International Workshop on Data Warehousing and OLAP, pp. 57–64 (2012)
10. Hilprecht, B., Binnig, C., Röhm, U.: Towards learning a partitioning advisor with deep reinforcement learning. In: aiDM Workshop, pp. 1–4 (2019)
11. Jindal, A., Karanasos, K., Rao, S., Patel, H.: Selecting subexpressions to materialize at datacenter scale. Proc. VLDB Endow. **11**(7), 800–812 (2018)
12. Karanasos, K., et al.: Dynamically optimizing queries over large scale data platforms. In: ACM SIGMOD, pp. 943–954 (2014)
13. Kipf, A., Kipf, T., Radke, B., Leis, V., Boncz, P., Kemper, A.: Learned cardinalities: estimating correlated joins with deep learning. arXiv preprint arXiv:1809.00677 (2018)
14. Kocsis, Z.A., Drake, J.H., Carson, D., Swan, J.: Automatic improvement of apache spark queries using semantics-preserving program reduction. In: GECCO, pp. 1141–1146 (2016)
15. Li, Y., Li, M., Ding, L., Interlandi, M.: RIOS: runtime integrated optimizer for spark. In: ACM Symposium on Cloud Computing, pp. 275–287 (2018)
16. Lima, A.A.B., Furtado, C., Valduriez, P., Mattoso, M.: Parallel OLAP query processing in database clusters with data replication. DaPD **25**(1–2), 97–123 (2009)
17. Ma, L., Van Aken, D., Hefny, A., Mezerhane, G., Pavlo, A., Gordon, G.J.: Query-based workload forecasting for self-driving database management systems. In: ACM SIGMOD, pp. 631–645 (2018)
18. Nehme, R., Bruno, N.: Automated partitioning design in parallel database systems. In: ACM SIGMOD, pp. 1137–1148 (2011)
19. Quamar, A., Kumar, K.A., Deshpande, A.: SWORD: scalable workload-aware data placement for transactional workloads. In: EDBT, pp. 430–441 (2013)
20. Serafini, M., Taft, R., Elmore, A.J., Pavlo, A., Aboulnaga, A., Stonebraker, M.: Clay: fine-grained adaptive partitioning for general database schemas. VLDB Endow. **10**(4), 445–456 (2016)
21. Stöhr, T., Märtens, H., Rahm, E.: Multi-dimensional database allocation for parallel data warehouses. In: VLDB, pp. 273–284 (2000)
22. Taft, R., et al.: E-store: fine-grained elastic partitioning for distributed transaction processing systems. VLDB Endow. **8**(3), 245–256 (2014)
23. Zhang, T., Tomasic, A., Sheng, Y., Pavlo, A.: Performance of OLTP via intelligent scheduling. In: ICDE, pp. 1288–1291 (2018)
24. Zhang, W., Kim, J., Ross, K.A., Sedlar, E., Stadler, L.: Adaptive code generation for data-intensive analytics. Proc. VLDB Endow. **14**(6), 929–942 (2021)
25. Zilio, D.C., et al.: Db2 design advisor: integrated automatic physical database design. In: VLDB, pp. 1087–1097 (2004)

Selecting Subexpressions to Materialize for Dynamic Large-Scale Workloads

Mustapha Chaba Mouna[1], Ladjel Bellatreche[2(✉)], and Narhimene Boustia[1]

[1] LRDSI Laboratory, Faculty of Science, University Blida 1, Soumaa, Blida, Algeria
[2] LIAS/ISAE-ENSMA, Poitiers, France
bellatreche@ensma.fr

Abstract. The nature of analytical queries executed either inside or outside of a DBMS increases the redundant computations due to the presence of common query sub-expressions. More recently, a few largely industry-led studies have focussed on the problem of identifying beneficial sub-expressions for large-scale workloads running outside of a DBMS for materialization purposes. However, these works have unfortunately ignored the large-scale workloads running inside of a DBMS. To align them in terms of the materialization of beneficial sub-expressions for dynamic large-scale workloads, we propose a pro-active approach that uses hypergraphs. These structures exploit cost models towards capturing the common query sub-expressions and materializing the most beneficial ones. Our approach is accompanied by a strategy, which orients the first δ queries to the offline phase that selects their appropriate views. To augment the benefit and sharing of the selected views, the initial δ queries may be scheduled. The online phase manages the pool of views obtained by the first phase by adding/dropping views to optimize new incoming queries. We conducted extensive experiments to evaluate the efficiency of our proposal as well as its cost-effective integration in a commercial DBMS.

Keywords: Query interaction · Dynamic hypergraphs · View selection

1 Introduction

Repeated computations on redundant data are pervasively invoked in modern data storage systems designed to intensively process Big Data Analytics Workflows. The nature of analytical queries increases the recurrent computations [22]. The management of these overlapping computations arouses interest in the three major generations of data processing: OTLP (e.g., multi query optimization [21]), OLAP (e.g., materialized view selection [7]), and RTAP (e.g., parallel processing of massive datasets [6]). This phenomenon continues to grow. We would like to highlight an important finding reported in [26] that more than 18% of queries of *real-world workloads of some Alibaba Cloud projects* are redundant; this represents a high number from a sharing perspective. Two

© Springer Nature Switzerland AG 2021
M. Golfarelli et al. (Eds.): DaWaK 2021, LNCS 12925, pp. 39–51, 2021.
https://doi.org/10.1007/978-3-030-86534-4_4

queries are considered to be redundant if they are similar or if they share common sub-expression(s). Identifying highly beneficial query sub-expressions to reduce computations is known as the problem of Sub-expression Selection (PSS) [8]. PSS is strongly connected to various important problems that we can divide into two main categories: **(i)** problems managed inside of a DBMS such as multi-query optimization [21], materialized view selection [7] and query rewriting [1] and **(ii)** problems managed outside of a DBMS like optimizing massive data jobs [6] and Cloud computing tasks [22].

Materialized view selection problem (VSP) is one of the most important problems that uses query sub-expressions. It aims at selecting sub-expressions that offer high benefit depending on the used objective functions for the whole workload, while respecting constraints such as storage budget [17], maintenance overhead [7] and query rewriting cost [1]. VSP has been widely studied in the literature, especially during the 1990–2000 period, where more than 180 papers with "view selection" in the title can be found in googlescholar.com. Elegant solutions have been implemented in leading DBMSs. Unfortunately, they usually consider only a few tens of static queries [26]. This situation is inadequate w.r.t. the requirements of today's analytical applications, where dynamic large-scale workloads involving redundant sub-expressions are considered. In parallel, research efforts conducted in managing massive user jobs and cloud queries have largely succeeded in dealing with large-scale workloads. For instance, $BIGSUBS$ system [8] has been able to manage workloads ranging from 2000 to 50,000 sub-expressions with the goal to materialize some of them. Based on this assessment, we figure out that the defenders of outside the DBMS outperform the defenders of inside the DBMSs in managing large-scale workloads. This situation has to be equilibrated among both of these defenders. One of the leading DBMS vendors presented a paper in VLDB'2020 [1] that tackles VSP under three properties: only a few of the views are selected, with reasonable sizes, that optimize joins and groupings, and can rewrite a substantial number of current and future workload queries. The authors of this paper promote the idea of developing proactive approaches for VSP, but they do not develop it adequately to address its complexity and implementation. Another point concerns the use of 650 queries running on a star schema [1] to evaluate the fixed requirements. This represents a significant number compared to previous studies conducted by academia and industry.

With the motivation of selecting views for dynamic large-scale workloads inside of a DBMS, in this paper, we propose to deal with the PSS to capture the most beneficial common sub-expressions. To do so, we first start by analyzing the existing studies dealing with massive user jobs and Cloud queries to identify the key features allowing them in managing large-scale workloads. We determined that they use graph structures to capture the common query sub-expressions [8,26]. In [26], the problem of sub-expression selection is modeled as a bipartite labeling problem and then design an Integer Linear Programming to solve them, which aims at maximizing the utility by selecting optimal common sub-expressions to be materialized. Jindal et al. [8] proposed a vertex-centric programming algorithm to iteratively choose in parallel sub-expressions to materialize for large-scale user jobs. In inside of DBMS findings, to the best of our knowledge, the sole work that uses hypergraphs to capture common query sub-expressions has been proposed in [4]. The usage of hypergraphs is motivated by their capacity of managing complex problems and their partitioning tools. Certainly, these studies manage large-scale workloads, but fail to consider their dynamic aspect. The diversity of data analysts and the changing business directives make the query workload more dynamic [19]. This dynamism has contributed to increase the research studies on self-tuning and autonomous databases [1,14]. Several reactive approaches such as *DynaMat* [11], *WATCH-MAN* [20], and MQT advisor [16] for dynamic materialized view selection have been proposed. The main characteristic of these approaches is that they react to transient usage, rather than relying purely on a historical workload [15].

In this paper, we propose a pro-active approach to capture most beneficial common sub-expressions and materializing them by respecting the storage constraint. The maintenance constraint is not studied in this paper. Since our queries arrive dynamically, based on a threshold, the first δ incoming queries are routed to the offline phase. Their common sub-expressions are captured by a hypergraph. Cost models are used to compute the benefit of each sub-expression. To augment the sharing of views, these queries may be scheduled. The online module exploits the findings of the first module to optimize the coming queries through three actions: (i) the use of the already selected views, (ii) the elimination of less beneficial ones, and (iii) the selection of new ones as in the offline phase.

The paper is organized as follows: Sect. 2 gives our related work. Section 3 presents our background. Section 4 describes in details our proactive approach. Section 5 presents our validation. Section 6 concludes the paper.

2   Related Work

The problem of subexpression selection is the centrality of several important Inside and Outside DBMS problems such as VSP, multi-query optimization, query scheduling, processing of user jobs over massive datasets, and the estimation of the cardinality of these subexpressions.

With regard to VSP, several formalizations have been proposed including different non-functional requirements and constraints. These formalizations have been accompanied by a large variety of algorithms [12], most of them consider a *small* set of static queries. This contradicts the ad-hoc nature of analytical queries [8]. Yang et al. [24] are considered as the pioneer that consider simultaneously the problems of selection subexpressions and materialized views. They proposed a greedy and 0–1 integer programming solutions to select the most beneficial expressions by considering only few queries. To deal with the dynamic nature of analytical queries a couple of systems have been proposed such as Dynamat system [11]. It monitors permanently the incoming queries and uses a pool to store the best set of materialized views based on a goodness metric and subject to space and update time constraints. These systems are not designed to anticipate future queries and use their common subexpressions to materialize them.

Multi-query optimization problem has been studied in the inside and outside DBMS contexts [18]. It consists in finding the best merging of query plans that optimizes the whole workload. Contrary to VSP, this problem does not consider constraints such as storage space. It is proven as NP-hard in [23]. Optimal algorithms have been proposed considering reasonable workloads [23]. To deal with large workloads involving huge common subexpressions, recently, approximate algorithms have been proposed in time quadratic to the number of common subexpressions [10]. The work proposed in [24] is the pioneer in the context of data warehouse, where it considers the results of a multi-query optimization solution to select materialized views. Although its novelty at that time, their work considers less than ten queries. To counter the scalability issue, hypergraphs have been proposed in [3,13] for coupling these two problems by considering static workloads. To deal with the dynamic aspect of workloads, a couple of research efforts consider query scheduling. In [16], a dynamic formalization of VSP is given by considering query scheduling policy based on a *genetic algorithm*. Their solution is strongly dependent on the DB2 advisor and manages just over a hundred queries. SLEMAS system proposed in [3] consider query scheduling when materializing a large set of static queries,

The above ideas have been revisited and leveraged to the context of Cloud query processing and user jobs for massive datasets [22] by considering graph structures to manage large scale workloads. An interesting recent direction has to be highlighted that consists in using deep learning and reinforcement learning to estimate the benefit of common expressions in optimizing the whole workload, their cardinality [26], and to select join order of workload queries that impact the selection of these common subexpressions [25].

3   Background

In this section, we present fundamental notions and definitions that facilitate the description of our proposal.

Definition 1. *A hypergraph $H = (V, E)$ is defined as a set of vertices V (nodes) and a set of hyper-edges E, where every hyper-edge connects a non-empty subset of nodes [5]. A hypergraph can be represented by an incidence matrix, where its rows and columns represent respectively vertices and hyperedges. The $(i, j)th$ value in the matrix is equal to 1 if vertex v_i is connected in the hyperedge e_j, and is 0 otherwise.*

Definition 2. *The degree of a vertex $v_i \in V$, denoted by $d(v_i)$ represents the number of distinct hyper-edges in E that connect v_i.*

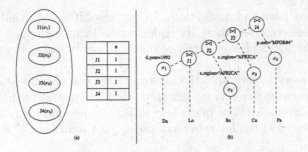

Fig. 1. From a SQL query to hypergraph (a) and Query tree (b)

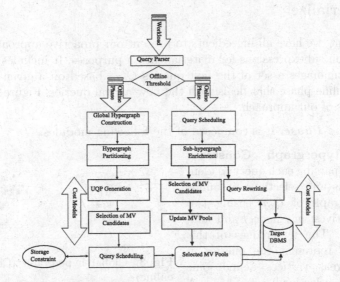

Fig. 2. The global architecture of our proposal

Hypergraphs are a powerful structure to represent the relationships that may exist in complex systems such as query processing [2]. An analytical query can represented by a hypergraph with one hyperedge representing the query and the nodes define the query operations (joins and selections) [3]. To illustrate this point, let us consider the following analytical query Q defined on the star schema benchmark (SSB)[1] that contains a fact table *Lineorder* (LO) and four dimension tables *Customer* (CU), *Supplier* (SU), *Part* (PA), and *Dates* (DA).

```
Q: SELECT  d_year, s_nation, p_category, sum(lo_revenue - lo_supplycost) as profit
FROM  DATES Da, CUSTOMER Cu, SUPPLIER Su, PART Pa, LINEORDER Lo
WHERE  lo_orderdate = d_datekey  (J1) and lo_suppkey = s_suppkey  (J2)
AND lo_custkey = c_custkey  (J3) and lo_PARTkey = p_PARTkey  (J4)
AND c_region = 'AFRICA' and s_region  'AFRICA' and d_year = 1992 and p_mfgr = 'MFGR#4'
GROUP BY  d_year, s_nation, p_category ORDER BY  d_year, s_nation, p_category;
```

---

[1] http://www.cs.umb.edu/~poneil/StarSchemaB.pdf.

Q contains 4 joins and 4 selections. In our case, if a join is subject to filter condition(s) (selection(s)), then, its node has to include these selections. The hypergraph and its incidence matrix of our SQL analytical query are shown in Fig. 1.

From the query hypergraph, the algebraic tree can also be generated if the type of join processing tree (e.g., left deep, right deep, and bush) and the join order are a priori known. The query tree in Fig. 1(b) is obtained by considering left deep strategy and the following join order ($DA \bowtie LO \bowtie SU \bowtie CU \bowtie PA$).

4   Proactive Selection of Common Subexpression to Materialize

At this point, we have all ingredients to present our proactive approach to capture common subexpressions for materialization purposes. It includes an offline phase that manages a set of the first queries Q_{off} based on a given threshold δ and an online phase that deals with the new arrival queries. Figure 2 gives all components of our approach.

The Offline Phase. It is composed of the following modules.

Global Hypergraph Construction: After parsing each query to identify its joins and selections, it builds the hypergraph of Q_{off} using two main primitives: *add-node()* and *add-hyperedge()*. The hyperedges of this hypergraph represent the queries in Q_{off}, whereas vertices define joins (and their selections) of Q_{off}. A hyperedge e_i connecting a set of vertices

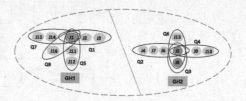

Fig. 3. Global hypergraph (Color figure online)

corresponds to a join node that participates in the execution of a query $Q_j \in Q_{off}$. The construction of this global hypergraph extends this for a single query, by maximizing sharing among nodes.

To illustrate this module, let us consider 8 queries[2]. The obtained global hypergraph contains two disjoint components including respectively GH_1 : (Q_1, Q_5, Q_7, Q_8) and GH_2 : (Q_2, Q_3, Q_4, Q_6) as shown in Fig. 3. Having a hypergraph with two components is explained by the fact that several queries do not share selections. Our hypergraph contains three nodes J_1, J_5 and J_8 with a high sharing that will be candidates for materialization.

Hypergraph Partitioning: To facilitate the management of our global hypergraph, its partitioning into sub-hypergraphs is required. This partitioning has to *maximize the query interaction* in each sub-hypergraph and <u>minimize their connections</u>. To perform this partitioning, we adapt the hMeTiS

---

[2] available at: https://drive.google.com/file/d/1s6StU4iXdQZlDJ0b9shoMAqdf9my CviI/view?usp=sharing.

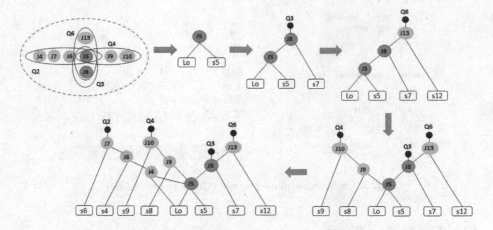

Fig. 4. From hypergraph to unified query plan

algorithm [9]. Contrary to hMeTiS, where the number of final partitions is known in advance, in our case this number is unknown. The partitioning process of our hypergraph is performed using two rules: (R1) the set of vertices is divided if and only if the number of hyperedge cutting is null and (R2) Each bisection resulting from the hypergraph partitioning is divided in the same way until no more divisible hypergraph is found.

Unified Query Plan Generation: This module aims at generating the unified graph plan (UQP) for each sub-hypergraph. The transformation follows three main steps: (i) choosing a pivot node that corresponds to the node that offers the best benefit if it is materialized. The benefit of each node n_i is calculated as follows: $benefit(n_i) = (nbr\_use - 1) \times process\_cost(n_i) - constr(n_i)$, where nbr_use, $process\_cost(n_i)$ and $constr(n_i)$ represent respectively the number of queries that use the join node ni, the processing cost of n_i, and the construction cost of n_i. (ii) Transforming the pivot node from the hypergraph to the oriented graph. (iii) Removing the pivot node from the hypergraph. Figure 4 illustrates this step, where the J_5 and J_8 are selected as candidates' views.

Query Scheduling: To increase the benefit and reusing of materialized views, we propose to reschedule the queries in Q_{off}. The scheduler is based on maximizing the benefit of reusing nodes, so the order is guided by nodes. To clarify this module, let us consider the hypergraph component $GH2$ illustrated in Fig. 3, in which 4 queries are involved. Two join nodes are candidates for materialization: J_5 and J_8 represented by green nodes. The first step of our process is to order the join nodes according to their benefits. Secondly, each query is assigned to a weight calculated by summing the benefit of its used nodes with maximal benefit. Finally, we order queries according to their weights. Figure 5 illustrates the scheduling process.

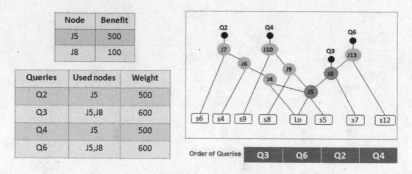

Fig. 5. Query scheduling process (Color figure online)

4.1   The Online Phase

Two main modules characterized this phase.

- *Enrichment of our Hypergraph.* This enrichment is performed by placing the incoming queries in their appropriate components and materializing the most shared joins dynamically identified. In order to increase the reuse of materialized views already selected and stored in the pool of a sub-hypergraph SHG_i^t, we define a metric for the incoming query Q^t called $Nbr\_Shared\_Views(Q^t, SHG_i^t)$ defined by the following equation:

$$Nbr\_Shared\_Views(Q^t, SHG_i^t) = |Joins(Q^t) \bigcap Pool^{SGH_i^t}| \qquad (1)$$

  Finally, Q^t is placed in the component that maximizes this metric. In the case, where Q^t does not share any materialized views with the existing sub-hypergraph, Q^t is placed in the component that shares with it the maximum of join nodes.
- *Dynamic Re-Selection of Materialized Views.* The dynamic and continuous arrival of queries produce a massive number of views to materialize which violates the storage space constraint. To cope with this issue, we put forward a strategy that consists in materializing the most shared joins that can be used by future incoming queries. To do that, we repeat the following process at each arrival of a new query to a component until the saturation of the fixed storage space: At the arrival of a new query Q^t and its placement to its subhypergraph SGH_i^t, we calculate the benefit of each join of Q^t using the benefit metric. Afterward, we check if there are joins having a positive benefit. If the storage space is saturated, we drop the materialized views having the least benefit in the pool of this subhypergraph and we replace them by materializing beneficial joins of the current query. To illustrate our dynamic strategy for managing ad-hoc and dynamic queries, an example is given in the Fig. 6 by considering the same query workload and hypergraph considered in the previous examples.

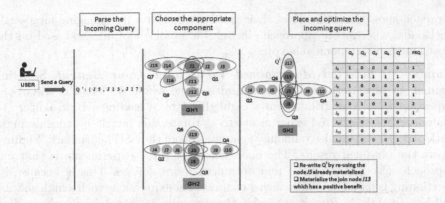

Fig. 6. The optimization process of a new incoming query

5 Experimental Study

This section shows the results of our experiments that we conducted to evaluate the efficiency and effectiveness of our proposal. We use the following environment: a server with E5-2690V2, 3 GHz processor, 24 GB of main memory, and 1 TB of the hard disk. The SBB schema is deployed in Oracle 12c DBMS. Our proposal is compared against two existing approaches: Yang [24] and Phan [16], since they tackle the same problem.

Scalability Issue. In this experiment, we study the capacity of three approaches in dealing with a large-scale workload. To do so, we vary the size of our workloads from 1000 to 5000 queries. The obtained results are depicted in Fig. 7. They show the scalability of our approach and the limitation of Yang's algorithm in managing large-scale workloads. The use of hypergraphs and the adaptation of hMeTis partitioning are some key factors of this scalability. The genetic algorithm proposed by [16] explores the search space of $n!$ representing the number of permutations to find the best scheduling. The results obtained by Phan's

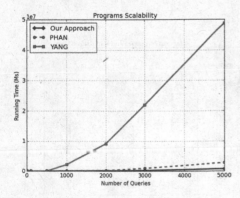

Fig. 7. Scalability comparison

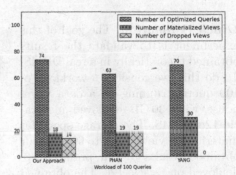

Fig. 8. Number of optimized queries and selected views

approach show the difficulty of their genetic algorithm in managing large-scale workloads, whereas our approach schedules a small set of queries based on the most beneficial common subexpressions.

Number of Selected Materialized Views and Their Benefit. Selecting a small set of materialized views optimizing the whole workload is one of the important quality metrics recently highlighted by a leading DBMS editor [1]. Therefore, we conducted experiments to evaluate this metric by considering a workload of 1 000 queries randomly generated from the SSB benchmark. Figure 8 shows the obtained results. The main lesson of these experiments is that our approach selects fewer views and optimizes more queries. This is because its functioning is based on materializing common subexpressions with a high sharing and benefit. Like in Phan's approach, these results are obtained thanks to drop unnecessary views and scheduling queries in the offline phase.

Quality of Our Cost Models. Our proposal is based on one important component representing mathematical cost models related to query processing cost with/without views, the construction cost of materialized views, and storage constraints. These cost models are found in [3]. The obtained results are reported in Fig. 9 which confirms the results obtained by our used DBMS.

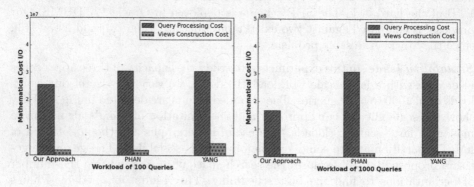

Fig. 9. Processing/maintenance costs of three approaches.

Oracle Validation. The goal of this experiment is to validate the results obtained theoretically in a real DBMS. To do this, we consider a workload of 100 queries running on a data warehouse with 30 GB deployed in Oracle 12c DBMS. The storage space for materialized views is set to 60 GB. Our obtained results by our proposal are deployed on Oracle 12c. The obtained results in terms of execution time

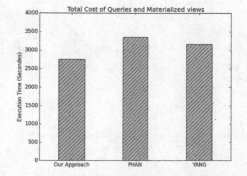

Fig. 10. Oracle validation

(in secs) are reported in Fig. 10. We observe that our approach outperforms the others. The obtained results coincide with those obtained theoretically. Both confirm the efficiency of our approach.

Impact of the Threshold δ on Query Processing Cost. In this experiment, we first evaluate theoretically and in Oracle 12c the overall processing cost of a workload of 100 queries by varying the offline threshold value. We observe in Fig. 11a, 11b an inverse relationship between the threshold values and the processing cost. At each increase in the threshold value, the processing cost decreased until the threshold value reaches 70 which represents the stability point of the processing cost optimization.

Our Wizard. Based on our findings, we have developed a tool, called ProRes Fig. 12 inspired by the well-known commercial advisors allows assisting DBA in their tasks when selecting materialized views for large-scale workloads. ProRes is developed using Java and integrates all our cited phases. A video showing all functionalities of our ProRes is available at: https://www.youtube.com/watch?v=_1p0RNkg9dw&ab_channel=MustaphaChabaMouna.

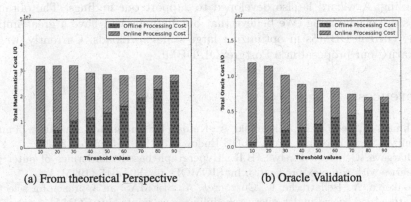

(a) From theoretical Perspective   (b) Oracle Validation

Fig. 11. The effect of the threshold

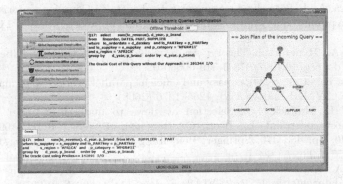

Fig. 12. An example of functioning of ProRes.

6   Conclusion

In this paper, we attempt to convince the defenders of inside DBMS of the urgent need of augmenting this nice technology by the recent results obtained by the defenders of outside DBMS in managing large-scale jobs and Cloud queries. Today, analytical workloads involve very large sets of dynamic and interacted queries. Several companies are still using traditional DBMSs that have to manage and optimize these workloads. To do so, we propose to study simultaneously the two correlated problems that consist in selecting common subexpressions of queries and materialized views. We propose the usage of hypergraphs to capture the common subexpressions due to their scalability offered by their partitioning mechanism. A proactive approach composed of two phases: an offline phase for optimizing the first queries δ based on a given threshold. The other coming queries are managed by the online phase that exploits the selected views of the offline one. Our approach keeps only beneficial views and schedules queries to improve the benefit of selected views. Our approach is compared against the pioneering work that studied the problems of selecting common expressions and materialized views and another one dealing with materialized views and query scheduling. A wizard is also developed to support our findings. The obtained results are encouraging. We believe that our findings will have a great impact on the current DBMSs in optimizing large-scale workloads. Currently, we are integrating our proposal into PostgreSQL DBMS.

References

1. Ahmed, R., Bello, R.G., Witkowski, A., Kumar, P.: Automated generation of materialized views in oracle. Proc. VLDB Endow. **13**(12), 3046–3058 (2020)
2. Bhargava, G., Goel, P., Iyer, B.R.: Hypergraph based reorderings of outer join queries with complex predicates. In: SIGMOD, pp. 304–315 (1995)
3. Boukorca, A., Bellatreche, L., Cuzzocrea, A.: SLEMAS: an approach for selecting materialized views under query scheduling constraints. In: COMAD, pp. 66–73 (2014)
4. Boukorca, A., Bellatreche, L., Senouci, S.B., Faget, Z.: Coupling materialized view selection to multi query optimization: hyper graph approach. IJDWM **11**(2), 62–84 (2015)
5. Bretto, A.: Hypergraph Theory: An Introduction. Springer, Heidelberg (2013). https://doi.org/10.1007/978-3-319-00080-0
6. Chaiken, R., et al.: SCOPE: easy and efficient parallel processing of massive data sets. Proc. VLDB Endow. **1**(2), 1265–1276 (2008)
7. Gupta, H., Mumick, I.S.: Selection of views to materialize under a maintenance cost constraint. In: Beeri, C., Buneman, P. (eds.) ICDT 1999. LNCS, vol. 1540, pp. 453–470. Springer, Heidelberg (1999). https://doi.org/10.1007/3-540-49257-7_28
8. Jindal, A., Karanasos, K., Rao, S., Patel, H.: Selecting subexpressions to materialize at datacenter scale. Proc. VLDB Endow. **11**(7), 800–812 (2018)
9. Karypis, G., Aggarwal, R., Kumar, V., Shekhar, S.: Multilevel hypergraph partitioning: application in VLSI domain, pp. 526–529 (1997)

10. Kathuria, T., Sudarshan, S.: Efficient and provable multi-query optimization. In: PODS, pp. 53–67 (2017)
11. Kotidis, Y., Roussopoulos, N.: DynaMat: a dynamic view management system for data warehouses. In: SIGMOD, pp. 371–382 (1999)
12. Mami, I., Bellahsene, Z.: A survey of view selection methods. ACM SIGMOD Rec. **41**(1), 20–29 (2012)
13. Mouna, M.C., Bellatreche, L., Boustia, N.: HYRAQ: optimizing large-scale analytical queries through dynamic hypergraphs. In: IDEAS, pp. 17:1–17:10 (2020)
14. Pavlo, A., et al.: External vs. internal: an essay on machine learning agents for autonomous database management systems. IEEE Data Eng. Bull. **42**(2), 32–46 (2019)
15. Perez, L.L., Jermaine, C.M.: History-aware query optimization with materialized intermediate views. In: ICDE, pp. 520–531 (2014)
16. Phan, T., Li, W.: Dynamic materialization of query views for data warehouse workloads. In: ICDE, pp. 436–445 (2008)
17. Roukh, A., Bellatreche, L., Bouarar, S., Boukorca, A.: Eco-physic: eco-physical design initiative for very large databases. Inf. Syst. **68**, 44–63 (2017)
18. Roy, P., Sudarshan, S.: Multi-query optimization. In: Liu, L., Özsu, M.T. (eds.) Encyclopedia of Database Systems. Springer, New York (2018). https://doi.org/10.1007/978-1-4614-8265-9_239
19. Savva, F., Anagnostopoulos, C., Triantafillou, P.: Adaptive learning of aggregate analytics under dynamic workloads. Future Gener. Comput. Syst. **109**, 317–330 (2020)
20. Scheuermann, P., Shim, J., Vingralek, R.: WATCHMAN: a data warehouse intelligent cache manager. In: VLDB, pp. 51–62 (1996)
21. Sellis, T.: Multiple-query optimization. ACM TODS **13**(1), 23–52 (1988)
22. Silva, Y.N., Larson, P., Zhou, J.: Exploiting common subexpressions for cloud query processing. In: ICDE, pp. 1337–1348 (2012)
23. Timos, K., Sellis, S.G.: On the multiple query optimization problem. IEEE Trans. Knowl. Data Eng. **2**, 262–266 (1990)
24. Yang, J., Karlapalem, K., Li, Q.: Algorithms for materialized view design in data warehousing environment. In: VLDB, pp. 136–145 (1997)
25. Yu, X., Li, G., Chai, C., Tang, N.: Reinforcement learning with tree-LSTM for join order selection. In: ICDE, pp. 1297–1308 (2020)
26. Yuan, H., Li, G., Feng, L., Sun, J., Han, Y.: Automatic view generation with deep learning and reinforcement learning. In: ICDE, pp. 1501–1512 (2020)

Prediction Techniques

A Chain Composite Item Recommender for Lifelong Pathways

Alexandre Chanson[1], Thomas Devogele[1], Nicolas Labroche[1],
Patrick Marcel[1(✉)], Nicolas Ringuet[1,2], and Vincent T'Kindt[1]

[1] University of Tours, Tours, France
{alexandre.chanson,thomas.devogele,nicolas.labroche,patrick.marcel,
vincent.tkindt}@univ-tours.fr, nicolas.r@neolink.link
[2] Neolink, Blois, France

Abstract. This work addresses the problem of recommending lifelong
pathways, i.e., sequences of actions pertaining to health, social or professional aspects, for fulfilling a personal lifelong project. This problem
raises some specific challenges, since the recommendation process is constrained by the user profile, the time they can devote to the actions in
the pathway, the obligation to smooth the learning curve of the user. We
model lifelong pathways as particular chain composite items and formalize the recommendation problem as a form of orienteering problem. We
adapt classical evaluation criteria for measuring the quality of the recommended pathways. We experiment with both artificial and real datasets,
showing our approach is a promising building block of an interactive
lifelong pathways recommender system.

Keywords: Chain composite item recommendation · Orienteering
problem

1 Introduction

We consider in this work the problem of building a recommendation system to
support social actors and beneficiary users in the interactive co-construction of
a personal lifelong project, for example the assistance of job seekers, or elderly
home support. Such a system would help different social actors to interact for
building a personal project for beneficiaries from a very large set of possible
actions pertaining to health, social or professional aspects. Such actions have
an intrinsic cost, they should complement well one another and be relevant
for the beneficiary, and doable in a period of time suitable for the beneficiary.
We denote the long-term sequence of actions proposed to beneficiaries as their
lifelong pathways. Recommending lifelong pathways raises many challenges: (i)
formalizing the recommender output, i.e., lifelong pathways, (ii) formalizing the
problem of computing pathways adapted to beneficiary users, (iii) determining
the quality of the recommended pathways.

Funded by ANRT CIFRE 2020/0731.

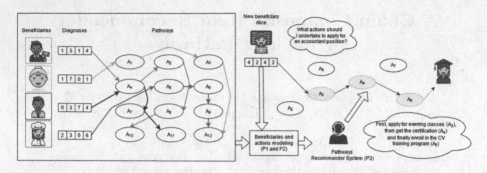

Fig. 1. Pathway Recommender System (PRS). P_1, P_2 and P_3 relate to the formal problems defined in Sect. 2.

Let's consider the example of Alice as pictured in Fig. 1 who wants to know how to achieve her personal long-term goal (apply to an accountant position). Alice has already achieved a diagnosis of her current situation, similar to previous beneficiaries for which diagnoses and pathways were observed beforehand. The objective of the pathways recommender system (PRS for short) is to determine from there what would be the most relevant next actions for Alice to undertake to achieve her long-term goal. This is a complex problem as it is heavily constrained: Alice has a limited time budget, and has to follow a certain learning curve so that each new action can easily build over the previously achieved actions. As a consequence, PRS has to estimate from previously observed beneficiaries, diagnoses, actions and their respective pathways, what would be the cost and the relevance for Alice to undertake an action. Similarly, the PRS needs to estimate a distance between actions to smooth the learning curve of Alice. Finally, the number and the diversity of actions make this problem difficult to solve.

In this paper, we contribute with a first approach that sets the bases of lifelong pathways recommendation. We model the problem, study its complexity, propose a resolution and optimizations, as well as evaluation criteria. Consistently with previous works, we consider that recommending such composite items calls for non traditional recommendation approaches [2]. More precisely, we consider lifelong pathways as particular chain composite items (CCI) [8,10] built from atomic actions, that have to satisfy the classical CCI properties of maximality (in terms of relevance), validity (in terms of beneficiary budget), and compatibility (in terms of action composition). The recommendation problem is formalized as a form of orienteering problem [6,13], for which we implemented two approaches for finding exact and approximate solutions. We adapt classical evaluation criteria for measuring the quality of the recommended pathways. We experiment with both artificial and real datasets, showing our approach is a promising building block of an interactive lifelong pathways recommender system.

This paper is organized as follows. Section 2 gives the formal background and defines the problem, while Sect. 3 presents our approach. Section 4 details the experimental validation. Section 5 discusses related work, and Sect. 6 concludes and draws perspectives.

2 Formal Definitions

In this section, we briefly define the concepts on top of which we build our problem statement for the recommendation of lifelong pathways to beneficiaries.

2.1 Model

We consider a set A of atomic actions representing various personal, professional or health steps that can be undertaken by a beneficiary to further his life goal.

Each action $a \in A$ is associated with a cost noted $cost(a)$ that represents the action difficulty. Given two actions $a_1, a_2 \in A$, we consider a distance between them, noted $dist(a_1, a_2)$ that represents how smooth the progression of the beneficiary would be by pairing these two actions in a pathway.

In what follows, action costs and distances constitute the action profiles. Notably, we do not differentiate between types of actions, since this would be use case-dependent. However, they could be easily introduced by adapting cost and distance measure.

A beneficiary b is represented by a tuple of features $\overrightarrow{f}$, called a diagnosis (see Sect. 4.1 for examples of features).

A profile for b is a vector of relevance scores, one for each action. For a beneficiary b and an action a, we note $rel_b(a)$ the relevance of b in a. Given a beneficiary b and a set of actions A, a pathway p_b for b is a sequence of actions from A.

2.2 Problem Formulation and Complexity

Our recommendation problem has the following inputs: (i) a set of actions A for forming pathways, (ii) a set of former beneficiaries B represented by their diagnoses D and the set of pathways P they have undertook, formed by actions in A, (iii) a new beneficiary b with diagnosis $\overrightarrow{f}$. The problem consists of recommending for b a pathway p_b with actions of A, based on D, P and $\overrightarrow{f}$.

We decompose our recommendation problem in three sub-problems:

P_1 compute action profiles in terms of action cost and distance between actions,
P_2 compute a user profile in terms of relevance in actions,
P_3 compute recommendations of pathways.

Problem 1 (Action profiles). *Let B be a set of beneficiaries, A be a set of n actions and P be a set of pathways with actions in A, for the beneficiaries in B. The problem consists of computing a profile for each action a of A, i.e., $cost(a) \geq 0$ and, for all $a' \in A$, a metric $dist(a, a')$.*

Problem 2 (Beneficiary profile). *Let B be a set of beneficiaries, A be a set of n actions and P be a set of pathways with actions in A, for the beneficiaries in B. Let b be a new beneficiary not in B, with diagnosis $\vec{f}$. The problem consists of computing a profile for beneficiary b, i.e., a vector of relevance score for each action in A, noted $\vec{b} = \langle rel_b(a_1), \ldots, rel_b(a_n) \rangle$.*

Problem 3 (Pathway recommendation). *Let A be a set of n actions, each associated with a positive cost $cost(a_i)$ and a positive relevance score $rel_b(a_i)$. Each pair of action is associated with a distance $dist(a_i, a_j)$. Let b be a beneficiary with profile $\vec{b}$, and let t be a budget in terms of days left to the beneficiary for fulfilling the recommended pathway. The optimization problem consists in finding a sequence $p_b = \langle a_1, \ldots, a_m \rangle$ of actions, $a_i \in A$, without repetition, with $m \leq n$, such that:*

1. *maximize $\sum_{i=1}^{m} rel_b(a_i)$ (maximality)*
2. *minimize $\sum_{i=1}^{m-1} dist(a_i, a_{i+1})$ (compatibility)*

subject to

3. *$\sum_{i=1}^{m} cost(a_i) \leq t$ (validity).*

Complexity of P_3. Problem P_3 is strongly NP-hard [7] since the TSP can be reduced to it. Indeed, any instance of TSP can be turned into an instance of P_3 by assigning a cost of 0 to all cities along an relevance of 1 and leaving the distance unchanged, it is then trivial to show that an optimal solution to this P_3 instance is optimal for the original TSP instance. This result means that, unless $P = NP$, P_3 can only be solved to optimality by algorithms with a worst-case time complexity in $A^*(c^n)$, with c a positive root and n the size of A.

3   Problem Resolution

To remain close to what is empirically observed, problems P_1 and P_2 are mostly treated as learning problems, where distances between actions and the relevance of an action for the beneficiary are learned from the set of existing diagnoses and pathways, that were devised by professional social actors. The cost of an action can be more trivially extracted from past pathways available data.

3.1   Computing Action Profiles (P_1)

We model each action a in A as a vector of features, computed as the average of diagnoses of former beneficiaries who were recommended the action a. The cost of actions is simply fixed to the median time spent for each action, as reported in the set of past pathways P. The median is used due to its robustness to extreme values. Note that, should this information be unavailable, a machine learning based approach similar to the ones described next for distances and relevance computation, should be applied.

Learning distances between actions corresponds to a traditional metric learning problem. Classical approaches addressing this problem include (i) transformation to classification task [5,9,14], or (ii) weekly-supervised metric learning [15]. Regarding (i), distances are usually linear combinations of feature-wise distances between objects, obtained by fitting a linear classifier (e.g., SVM) over pairs of objects labeled positively if objects should be in the same group and negatively otherwise.

As to (ii), weekly-supervised metric learning approaches aim at defining a new space based on input pairwise Must-Link or Cannot-Link constraints that specify respectively if two points should be close or distant in the output space. A traditional method is to learn a generalized Mahalanobis distance [15], that is basically a generalization of the Euclidean distance defined as follows:

$$d_M(a_1, a_2) = \sqrt{(a_1 - a_2)^t M (a_1 - a_2)} \tag{1}$$

with $(a_1, a_2) \in A^2$ and where $M \succeq 0$ is a positive semi-definite matrix to ensure that d_M is a proper metric. In its simplest form, i.e. when $M = I$, d_M is a Euclidean distance and when M is diagonal, d_M changes the relative weights of each features in the computation of the distance. Finally, in the general case, it can express more complex scaling and relations between features. The method proposed in [15] relies on the following optimization problem to find the expression of M:

$$\min_M \sum_{(a_i, a_j) \in S} d_M^2(a_i, a_j) \tag{2}$$

$$\text{s.t.} \sum_{(a_i, a_j) \in D} d_M(a_i, a_j) \geq 1 \tag{3}$$

$$M \succeq 0 \tag{4}$$

where S (resp. D) is the set of constraints indicating that 2 actions should (resp. should not) appear in the same pathway in our case. Equation (3) is added to avoid the trivial solution $M = 0$. We followed this approach to learn the distances between actions.

3.2 Computing Beneficiary Profile (P_2)

Preliminaries. Recall from Sect. 2.2 that any beneficiary $b \in B$ is represented by a tuple of features $\vec{f} = \langle b_1, \ldots, b_f \rangle$. It is then possible to determine for each action $a \in A$ if beneficiary b has undertaken it. Thus, each action a can be represented as a tuple of f features $\vec{a}$, whose values are set as the average values observed for the beneficiaries $b \in B$ that undertook a in their respective pathways:

$$\vec{a} = \frac{1}{|B|} \sum_{b \in B} \langle b_1, \ldots, b_f \rangle \tag{5}$$

Collaborative Relevance Score. As can be seen in Fig. 1, the set of former pathways can be seen as a graph of the actions that were undertaken in the past. While this graph could have been used to extract action popularity scores for instance using Page Rank or centrality measures, which may be useful for recommending popular actions, our goal is to compute a different profile for each new beneficiary. As such, our problem is close to a cold start problem for a new user, for which we use a hybrid approach (in the sense of [11]), leveraging both the beneficiary diagnosis and the former diagnoses and pathways.

Precisely, we define the collaborative relevance of action $a \in A$ for beneficiary b by the probability $rel_b(a) = P(a|b)$ that beneficiary b undertakes the action a. We express this problem as a traditional binary classification task aiming to predict *True* if action a is relevant for b or *False* otherwise, based on the diagnosis $\overrightarrow{f}$ of b. We use a naive Bayes approach to obtain the expected probability values. By generalizing the prediction for a beneficiary based on the observations over past beneficiaries, this score accounts for a collaborative score of relevance.

3.3 Pathway Recommendation (P_3)

Problem P_3 can be seen as an extension of the orienteering problem [12] with service time. Such problem was already formulated in the Exploratory Data Analysis community, to find a series of relevant queries [6]. We choose to use the Mathematical Integer Programming (MIP) model proposed by [4] as we expect most instances to be small enough for an exact solution to be tractably found by a state-of-the-art solvers like CPLEX. Such model allows to trivially add linear constraints on a case by case basis, whenever needed. For efficiency purpose, the resolution of such problems classically reformulates the initial multi-objective problem into a single objective problem (the maximization of the relevance) and rewrites the last two remaining objectives (minimizing the time budget and the distance) as so-called epsilon constraints with upper bound on the time budget and the distance [4]. This reformulation benefits from single objective optimization mechanisms available in CPLEX.

For those larger instances of the problem that could not be handled by CPLEX, we also implemented a simple greedy algorithm to efficiently compute approximate solutions to P_3. Its principle is as follows. It starts by picking the most relevant action, and then adds subsequent actions by computing a score by dividing relevance by distance, and picking the actions achieving the best score. A windowing system is used to limit the number of comparisons (the larger the window, the greater the number of best relevant actions considered). Additionally, at each iteration, costly actions not respecting the remaining budget are pruned.

4 Tests

This section describes the experiments conducted to evaluate our approach. They answer 2 main questions: (i) how effective is our PRS approach to recommend lifelong pathways?, (ii) to which extent PRS scales to larger instances of the problem, i.e., to a larger population of possible actions?

Problems 1 and 2 where solved using Python, sklearn library providing an implementation of a Naive Bayes classifier, and metric-learn library providing the implementation of the algorithm proposed by [15]. The mathematical model relies on CPLEX 20.10 and is implemented in C++[1]. The greedy algorithm, GreedyPRS, is implemented in Java.

4.1 Effectiveness of PRS

Dataset. In this first test, we use a real dataset named "French RSA[2]" composed of 2812 user diagnoses and 56 actions, with 14 descriptive features that represent a set of beneficiary contextual information such as "need for housing assistance", "need for childcare", "health diagnosis", "searching for a job", etc. A few null values were replaced with the most frequent ones.

Evaluation Criteria. Our evaluation scheme uses classical evaluation metrics for recommender systems, namely precision and recall of the discovered composite items when compared to the real pathways that were undertaken by the beneficiaries. However, as the recommendation task is complex, we consider precision and recall at a certain similarity threshold. In what follows, we denote by $\sim$ a similarity function between actions. We consider that there is a match between a recommended action and an expected action from the pathway if their similarity exceeds a threshold that is a parameter of the precision and recall measures, similar to what is done in [1,5].

More formally, considering: (i) the set of all actions A, (ii) a real pathway P as a set of s actions $\{p_j\}_{j\in[1,s]}$ and, (iii) the set R of m recommended actions $\{r_i\}_{i\in[1,m]}$ produced by PRS, we define the True Positive set as $TP = \{r \in R \mid r \sim p, p \in P\}$, the False Positive set as $FP = R \setminus TP$ and the False Negative set as $FN = \{a \in A \setminus R \mid a \sim p, p \in P\}$. From these sets, it is possible to compute, $Recall = \frac{|TP|}{|TP|+|FN|}$, $Precision = \frac{|TP|}{|TP|+|FP|}$ and $F1\text{-}measure = 2\frac{Precision \cdot Recall}{Precision+Recall}$.

Methodology. As explained in Sect. 3.3, P_3 is handled by reformulating objectives as epsilon constraints for which an upper bound has to be provided. We experiment with several values for these bounds. Noticeably, we consider 3 thresholds for the cost constraints expressed as the run time of actions, respectively 207, 364 and 601 days as these are the average observed period of time for respectively, below 25 years old (y.o.), between 25 and 55 y.o. and above 55 y.o. to go back to work after an unemployment period[3]. Similarly, for the distance constraint we

---

[1] https://github.com/AlexChanson/Cplex-TAP

[2] RSA stands for Revenue de Solidarité Active and is French form of in work welfare benefit aimed at reducing the barrier to return to work.

[3] These are the most related official indicators that we found on the topic of social assistance giving hints how to set these thresholds in case of RSA social benefit.

consider 3 distance thresholds: 3, 4, and 5 that allows to run several actions in one recommended pathway.

Figure 2 provides histograms of the distributions of run time costs and distances which assess the choices for the aforementioned thresholds.

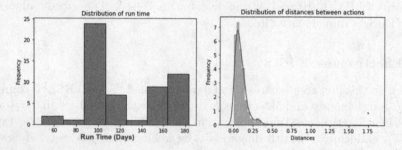

Fig. 2. Histograms of observed run time cost for the actions in the French RSA dataset (left) and histogram of distances between actions (right).

In order to assess the accuracy of our metric learning, we run a comparison with a simple Euclidean distance on all the previous scenarios. Finally, a 3-fold cross validation is set up to ensure that train and test sets are well separated and to produce average values.

Results. Figures 3, 4 and 5 summarize the main results of our experiments in terms of F-Measure at a given similarity threshold. Note that Figs. 4 and 5 represent different groupings of the same data, averaging distance and time, respectively. Results presented are averaged by the 3-fold cross validation process and standard deviation around each mean is represented as light color areas around the plots. Importantly, the similarity is expressed as a function of the distance $d(a_1, a_2)$ between actions as follows: $sim(a_1, a_2) = 1/(1 + d(a_1, a_2))$.

Figure 3 and 4 show the importance of the distance between actions in our method. Indeed, Fig. 3 presents results with a traditional Euclidean distance between actions, that does not benefit from external knowledge of which actions should be grouped together to better fit the observed pathways. Also, the lower the similarity threshold on theses figures, the easier it is to find a match between a recommender and an observed action, and the more likely the F-measure score is to be high. Consistently with this observation, both Figs. 3 and 4 show a decreasing F-Measure score with the increase in similarity threshold. However, our learned distance between action allows for a very high F-Measure until the similarity score is close to 1 while the traditional Euclidean distance falls for smaller similarity thresholds. Interestingly, in both cases, the F-Measure score when looking for a perfect match (similarity threshold = 1) are close to 0.3 but with slightly better performances for the tuned metric. Finally, the F-Measure score is higher in case of short-term pathways which is normal as it is easier to predict for short period of time and shorter sequences for which it is easier to improve the recall.

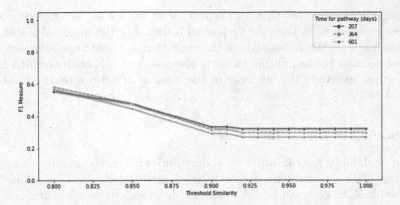

Fig. 3. F-measure scores computed for different similarity thresholds for 3 **run time costs** thresholds. Distance between action is set to an **Euclidean** distance.

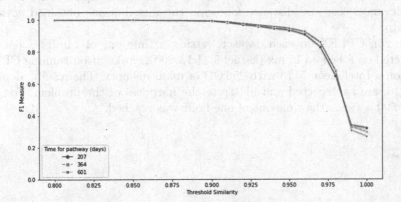

Fig. 4. F-measure scores computed for different similarity thresholds for 3 **run time costs** thresholds. Distance between action is learned as explained in Sect. 3.1.

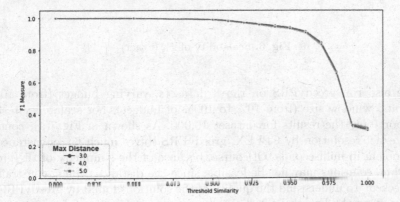

Fig. 5. F-measure scores computed for different similarity thresholds for 3 **distances** thresholds. Distance between action is learned as explained in Sect. 3.1.

Figures 4 and 5 show that Problem P_3 can be solved as efficiently if we consider constraints on the time budget as in Fig. 4 or the maximal distance as in Fig. 5. All in all, the formulation of the resolution of P_3 with epsilon constraints does not seem to be the limiting factor in the observed F-Measure results. In the next section, we study the influence of the number of actions on the resolution of Problem P_3.

4.2   Scaling to Larger Instances

For our scalability test, in order to understand what is the maximum number of actions PRS can handle, we generated 8 artificial datasets as follows. We first used a KDE-gaussian kernel to fit the distributions of relevance, cost and distance, respectively, with the samples obtained from the real dataset used for the effectiveness tests. We then generated 8 datasets of increasing sizes (100, 200, 300, 400, 500, 1000, 5000, 10,000), representing the inputs of P_3 by drawing relevance, cost and distance following the distributions obtained with the estimator.

We run CPLEX on each dataset, setting a time out of 1 h. This test was conducted on a Fedora Linux (kernel 5.11.13-200), workstation running CPLEX 20.10 on an Intel Xeon 5118 with 256 GB of main memory. The results, depicted in Fig. 6, are as expected and illustrate the hardness of the problem. For sizes above 500 actions, the time out of one hour was reached.

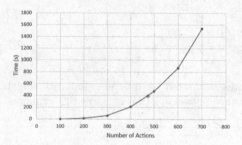

Fig. 6. Scalability of P_3 (exact)

We also run greedyPRS on the 8 datasets, varying budget (from 100 to 1000) and window size (from 10% to 100% of budget). For scalability testing, we report only the results for dataset 10,000. As shown in Fig. 7(a), contrary to the exact resolution by CPLEX, greedyPRS solves much larger instances of the problem in milliseconds. Of course, because of the simplicity of the greedy algorithm, solutions may not be feasible since the distance espilon-constraint is not checked. To understand the quality of the solutions found by greedyPRS, we also run it on the French RSA dataset, varying window size from 10 to 60. As illustrated in Fig. 7(b), the greedy approach maintains a reasonable F-measure in recommendation albeit not as good as PRS in our preliminary tests.

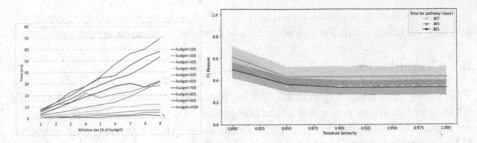

Fig. 7. Resolution of P_3 with GreedyPRS (approximate solution). Left: scalability. Right: F-measure.

5   Related Work

Our way to learn action profiles is similar to user intent discovery [5,9,14]. For instance, Guha et al. [9] discover user intent behind web searches, and obtain content relevant to users' long-term interests. They develop a classifier to determine whether two search queries address the same information need. This is formalized as an agglomerative clustering problem for which a similarity measure is learned over a set of descriptive features (the stemmed query words, top 10 web results for the queries, the stemmed words in the titles of clicked URL, etc.). A similar approach is used in [5] to discover BI user intents in BI queries.

Composite items (CIs) address complex information needs and are prevalent in problems where items should be bundled to be recommended together [2], like in task assignment in crowdsourcing or travel itinerary recommendation. CI formation is usually expressed as a constrained optimization problem, and different CI shapes require the specification of different constraints and optimizations. Our formulation of lifelong pathways is a particular case of chain shaped CIs [3,8,10], that are traditionally defined in terms of compatibility (e.g., geographic distance), validity (e.g., the total cost of an itinerary is within budget) and maximality (e.g., the itinerary should be of the highest value in terms of its POIs popularities), this last one often being used as the objective function. Retrieval of chain CIs is usually NP-hard, being reduced to TSP or orienteering problems, and has been addressed through greedy algorithm [3,10], dynamic programming or dedicated TSP strategies [8].

While being consistent with previous formulations, our formalization of lifelong pathway recommendations borrows from the Traveling Analyst Problem (TAP) [4,6]. This problem describes the computation of a sequence of interesting queries over a dataset, given a time budget on the query execution cost, and such that the distance between queries is minimized. TAP differs from the classical orienteering problem by adding a knapsack constraint to it. This formulation is close to that of [10], but the latter simplifies the problem by merging action cost and travel time budget.

In our case, similarly to [6], the distance we use has a semantics in itself and cannot be made analogous to a time or to a physical distance. Thus it must be considered as a separate constraint. Furthermore we are not specifying any starting or finish point for the sequence as the classical orienteering problem.

6   Conclusion

This paper introduces a system for lifelong pathways recommendation. We model lifelong pathways as particular chain composite items that are built from atomic actions, and whose retrieval is formalized as a form of orienteering problem. We experiment with both artificial and real datasets, showing our approach is a promising building block of an interactive lifelong pathways recommender system. Our short term perspectives consist of including in our model a multi-stakeholders context, to conciliate the objectives or constraints of beneficiaries, social actors and services providers, as well as mechanisms for the exploration of pathways, so that users can interact with the recommender. On the longer term, we plan to add an explanation mechanism of the recommendations to increase trust of stakeholders in the proposed pathways and to improve their acceptance.

References

1. Aligon, J., Gallinucci, E., Golfarelli, M., Marcel, P., Rizzi, S.: A collaborative filtering approach for recommending OLAP sessions. DSS **69**, 20–30 (2015)
2. Amer-Yahia, S., Roy, S.B.: Interactive exploration of composite items. In: EDBT, pp. 513–516 (2018)
3. Cao, X., Chen, L., Cong, G., Xiao, X.: Keyword-aware optimal route search. Proc. VLDB Endow. **5**(11), 1136–1147 (2012)
4. Chanson, A., Labroche, N., Marcel, P., T'Kindt, V.: The traveling analyst problem, orienteering applied to exploratory data analysis. In: ROADEF (2021)
5. Drushku, K., Aligon, J., Labroche, N., Marcel, P., Peralta, V.: Interest-based recommendations for business intelligence users. Inf. Syst. **86**, 79–93 (2019)
6. Chanson, A., et al.: The traveling analyst problem: definition and preliminary study. In: DOLAP, pp. 94–98 (2020)
7. Garey, M.R., Johnson, D.S.: "Strong" NP-completeness results: motivation, examples, and implications. J. ACM **25**(3), 499–508 (1978)
8. Gionis, A., Lappas, T., Pelechrinis, K., Terzi, E.: Customized tour recommendations in urban areas. In: WSDM, pp. 313–322 (2014)
9. Guha, R.V., Gupta, V., Raghunathan, V., Srikant, R.: User modeling for a personal assistant. In: WSDM, pp. 275–284 (2015)
10. Roy, S.B., Das, G., Amer-Yahia, S., Yu, C.: Interactive itinerary planning. In: ICDE, pp. 15–26 (2011)
11. Son, L.H.: Dealing with the new user cold-start problem in recommender systems: a comparative review. Inf. Syst. **58**, 87–104 (2016)
12. Tsiligirides, T.: Heuristic methods applied to orienteering. J. Oper. Res. Soc. **35**(9), 797–809 (1984)
13. Vansteenwegen, P., Gunawan, A.: Orienteering Problems. EURO Advanced Tutorials on Operational Research (2019)
14. Wang, H., Song, Y., Chang, M., He, X., White, R.W., Chu, W.: Learning to extract cross-session search tasks. In: WWW, pp. 1353–1364 (2013)
15. Xing, E.P., Ng, A.Y., Jordan, M.I., Russell, S.J.: Distance metric learning with application to clustering with side-information. In: NIPS, pp. 505–512 (2002)

Health Analytics on COVID-19 Data with Few-Shot Learning

Carson K. Leung$^{(\boxtimes)}$ ⓘ, Daryl L. X. Fung, and Calvin S. H. Hoi

University of Manitoba, Winnipeg, MB, Canada
kleung@cs.umanitoba.ca

Abstract. Although huge volumes of valuable data can be generated and collected at a rapid velocity from a wide variety of rich data sources, their availability may vary due to various factors. For example, in the competitive business world, huge volumes of transactional shopper market data may not be made available partially due to proprietary concerns. As another example, in the healthcare domain, huge volumes of medical data may not be made available partially due to privacy concerns. Sometimes, only limited volumes of privacy-preserving data are made available for research and/or other purposes. Embedded in these data is implicit, previously unknown and useful information and knowledge. Analyzing these data by data science models can be for social and economic good. For instance, health analytics of medical data and disease reports can lead to the discovery of useful information and knowledge about diseases such as the coronavirus disease 2019 (COVID-19). This knowledge helps users to get a better understanding of the disease, and thus to take parts in preventing, controlling and/or combating the disease. However, many existing data science models often require lots of historical data for training. To deal with the challenges of limited data, we present in this paper a data science system for health analytics on COVID-19 data. With few-shot learning, our system requires only limited data for training. Evaluation on real-life COVID-19 data—specifically, routine blood test results from potential Brazilian COVID-19 patients—demonstrates the effectiveness of our system in predicting and classifying of COVID-19 cases based on a small subset of features in limited volumes of COVID-19 data for health analytics.

Keywords: Data analytics · Knowledge discovery · Data mining · Machine learning · Few-shot learning (FSL) · Autoencoder · Data science · Healthcare data · Disease analytics · Coronavirus disease 2019 (COVID-19)

1 Introduction

Nowadays, technological advancements able the generation and collection of huge volumes of valuable data. For examples, data can be generated and collected at a rapid velocity from a wide variety of rich data sources such as healthcare

ⓒ Springer Nature Switzerland AG 2021
M. Golfarelli et al. (Eds.): DaWaK 2021, LNCS 12925, pp. 67–80, 2021.
https://doi.org/10.1007/978-3-030-86534-4_6

data and disease reports [1–4], imprecise and uncertain data [5–8], social networks [9–15], streaming data and time series [16–19], and transportation and urban data [20–23]. Embedded in these data is implicit, previously unknown and potentially useful information and knowledge that can be discovered by data science [24–26]. It makes good use of data analytics [27–31], data mining [32–38] (e.g., incorporating constraints [39, 40]), machine learning (ML) [41–43], mathematical and statistical modeling [44, 45], and/or visualization [46–48]. In addition to knowledge discovery, data science has also been used for data management; information retrieval [49] from the well-managed data; and data visualization and information interpretation for the retrieved information and discovered knowledge.

Health analytics on these data by data science can be for social good. For instance, analyzing and mining healthcare data and disease reports helps the discovery of useful information and knowledge about the disease such as *coronavirus disease 2019* (*COVID-19*), which was caused by severe acute respiratory syndrome-associated coronavirus 2 (SARS-CoV-2). This was reported to break out in late 2019, became a global pandemic in March 2020, and is still prevailing in 2021. Discovered information and knowledge helps prevent, detect, control and/or combat the disease. This, in turn, helps save patient life and improve quality of our life. Hence, it is useful to have a data analytics system for mining these health data.

In data analytics, it is common to train a prediction model with lots of data in many fields. Analysts look for trends, patterns and commonalities within and among samples. However, in the medical and health science field, samples can be expensive to obtain. Moreover, due to privacy concerns and other related issues, few samples may be made available for analyses. Take COVID-19 as an example of diseases. Although there have been more than 180 million confirmed COVID-19 cases and more than 3.9 million deaths worldwide (as of June 30, 2021)[1], the volume of *available* data may be limited partially due to privacy concerns (e.g., to privacy-preserving data publishing) and/or fast reporting of the information. This motivates our current work on designing a health analytics system that makes predictions based on limited data—via *few-shot learning* (*FSL*) with autoencoder.

Our *key contributions of this paper* include the design and development of our data science system for health analytics on COVID-19 data with FSL. Specifically, predictive models in our health analytics system can be trained by a few shots (i.e., a few samples of COVID-19 cases):

1. a model first trains the autoencoder (for reconstructed features based on the input features on the samples) before training the COVID-19 prediction outputs;
2. another model trains the autoencoder and the COVID-19 prediction outputs simultaneously (based on the input features); and
3. a third model trains the autoencoder and the COVID-19 prediction outputs simultaneously (based on the input features from *focused subsets* of samples).

---

[1] https://covid19.who.int/.

Evaluation results on real-life routine blood test results from potential Brazilian COVID-19 patients show that our system (which requires only a small training sample) makes more accurate predictions than many baseline ML algorithms (which require large training data).

We organize the remainder of this paper. Background and related works are provided in the next section. We then describe in Sect. 3 our health analytics system on COVID-19 data with FSL. Sections 4 and 5 show evaluation results and conclusions, respectively.

2   Background and Related Works

In this paper, our health analytics system analyzes COVID-19 data and makes prediction based on *FSL* [50], which is a ML technique that aims to learn from a limited number of examples in experience with supervised information for some classes of task. FSL has gain popularity in recent years. For instance, Snell et al. [51] applied the FSL with 5-shot modelling to classify images.

To conduct health analytics with the FSL, we adapt an *autoencoder* [52,53], which is an artificial neural network built for learning efficient data coding in an unsupervised mode. To reduce dimensionality, the autoencoder learns a data encoding (i.e., a representation of a set of data) by (a) training the network to ignore signal "noise" and (b) reconstructing from the reduced data encoding a representation as similar as possible to its original input. Usually, an autoencoder consists of two parts:

1. an encoder function $f{:}x \mapsto h$, and
2. a decoder function $g{:}h \mapsto r$ that represents a reconstruction r

where h is an internal representation or code in the hidden layer of the autoencoder. To copy its input x to its output r, the autoencoder maps x to r through h. Autoencoders have often been used in applications like anomaly detection, image processing, information retrieval, and popularity prediction. For medical applications, Ji et al. [54] used a deep autoencoder to infer associations between disease and microRNAs (miRNAs), which are non-coding RNAs. Similarly, Shi et al. [55] applied variational inference and graph autoencoder to learn a representation model for predicting long non-coding RNA (lncRNA)-disease associations. Gunduz [56] used variational autoencoders to reduce dimensionality for Parkinson's disease classification, but not for COVID-19 classification.

For COVID-19 classification and prediction, Rustam et al. [57] forecasted the number of new COVID-19 cases, death rate, and recovery rate. In contrast, we focus on predict the COVID-19 test outcome (i.e., COVID-19 positive or not). To predict the test outcome, Brinati et al. [58] applied random forest (RF) to routine blood test data collected from a hospital in Milan, Italy. Similarly, Xu et al. [59] applied a combination of traditional ML models (e.g., RF) and deep learning (e.g., convolutional neural networks (CNN)) to both lab testing data and computerized tomography (CT) scan images. Note that CT scan images are expensive to produce. Both RF and CNN usually require lots of data for training, which may not always be available. In contrast, our health analytics system trains our prediction models by a few samples while maintaining the prediction accuracy.

In one of our prediction model, we adapt *multitask learning (MTL)* [60,61]. It exploits commonalities and differences across multiple learning tasks to come up with a shared representation (for feature or representation learning), and solves these tasks in parallel by using the shared representation.

3   Our Health Analytics System

Our system is designed to conduct health analytics on a few shots of COVID-19 data to make accurate prediction of the outcome (i.e., COVID-19 positive or not). Here, we design three prediction models.

3.1   A Serial Prediction Model Based on Samples from the Dataset

The first prediction model is a serial one that makes prediction based on a few shots sampled from the entire set D of limited data. This *serHA(D)* prediction model is trained in serial in two key steps:

1. reconstruction of input features, and
2. prediction of COVID-19 outcome (i.e., COVID-19 positive or not).

To reconstruct input features, our *serHA(D)* model first builds an *autoencoder* with four fully connected layers. Specifically, the input layer takes n features $\{x_i\}_{i=1}^n$. With a hidden (first) layer, the encoder module f of the autoencoder maps the input features to produce m encoded features $\{h_j\}_{j=1}^m$ in the second layer. With another (third) hidden layer, the decoder module g of the autoencoder maps the m encoded features $\{h_j\}_{j=1}^m$ to produce n reconstructed input $\{r_i\}_{i=1}^n$ in the final (fourth) layer. See Fig. 1. During this first key step, the autoencoder aims to minimize the loss function L_R in the reconstruction of input features. This loss function L_R is computed as a mean absolute error:

$$L_R = \sum_{i=1}^n |r_i - x_i| \tag{1}$$

where (a) n is the number of trained data samples, (b) x_i is one of the n input features $\{x_i\}_{i=1}^n$ fed into (the encoder module of) the autoencoder, and (c) r_i is one of the n reconstructed features $\{r_i\}_{i=1}^n$ reconstructed by (the decoder module of) the autoencoder.

Once the autoencoder is well trained for reconstructing input features, our *serHA(D)* model then *freezes* the autoencoder and builds a prediction module—with a neural network having two fully connected layers—to make prediction on class labels (i.e., COVID-19 positive or not). Specifically, the input layer takes the m encoded features $\{h_j\}_{j=1}^m$ from the frozen autoencoder. With a hidden (first) layer, the prediction module maps the m encoded features $\{h_j\}_{j=1}^m$ to

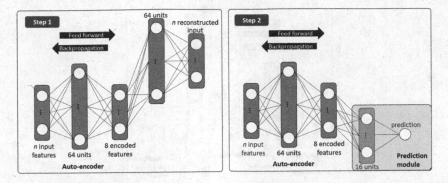

Fig. 1. Architecture for our *serHA(D)* prediction model in our health analytics system.

produce a single-label prediction y (i.e., COVID-19 positive or not) in the final (second) layer:

$$y = \begin{cases} 0 \text{ if predicted to be COVID-19 negative} \\ 1 \text{ if predicted to be COVID-19 positive} \end{cases} \tag{2}$$

Again, see Fig. 1. During this second key step, the prediction module aims to minimize the loss function L_P in the prediction of COVID-19 test outcome. This loss function L_P is computed as a binary cross-entropy loss:

$$L_P = \sum_{k=1}^{s} \left[y_k \log(y_k) + (1 - y_k) \log(1 - y_k) \right] \tag{3}$$

where s is the number of limited data samples per class label.

3.2 A Simultaneous Prediction Model Based on Samples from the Dataset

While the prediction model *serHA(D)* is trained *in serial* by training (a) the reconstruction of input features before (b) the prediction of COVID-19 outcome (i.e., COVID-19 positive or not), our second prediction model *simHA(D)* is trained *simultaneously* so that the resulting trained model also makes prediction based on a few shots sampled from the entire set D of limited data. In other words, this *simHA(D)* prediction model trains (a) the reconstruction of input features and (b) the prediction of COVID-19 outcome.

Similar to *serHA(D)*, our *simHA(D)* model reconstructs input features by building a 4 fully connected-layered autoencoder. However, unlike *serHA(D)*, our *simHA(D)* model does not freeze the autoencoder before building the prediction module. Instead, it makes prediction of class labels by building a 2 fully connected-layered neural network-based prediction module. Specifically, the input layer of the prediction module takes the m encoded features $\{h_j\}_{j=1}^{m}$ from the (*active*) autoencoder. See Fig. 2. During the simultaneous training of

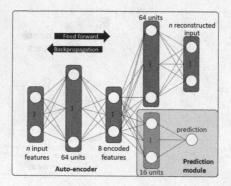

Fig. 2. Architecture for our $simHA(D)$ prediction model in our health analytics system.

both (a) the reconstruction of input features and (b) the prediction of COVID-19 outcome, our $simHA(D)$ model aims to minimize the weighted sum of the two corresponding loss functions L_R and L_P:

$$L_{sim} = \lambda_1 L_R + \lambda_2 L_P \tag{4}$$

where (a) users can specify weights $\lambda_1 < \lambda_2$ to emphasize the importance of COVID-19 prediction over input feature reconstruction, (b) L_R measures the loss in input feature reconstruction (Eq. (1)), and (c) L_P measures the loss in COVID-19 prediction (Eq. (3)).

3.3   Prediction Models Based on Samples from Focused Portions of the Dataset

Note that, in many healthcare data and disease reports, the number of records is usually small, partially due to difficulty in obtaining data and privacy concerns in making the data accessible. However, the number of available features within each record can be high. In real-life situations (e.g., COVID-19 situations), values may not be available in quite a number of these features. This may partially due to unavailability of resources to perform all the clinical diagnostic laboratory tests to obtain values for these features. The situations can be worsened in rural areas where resources can be further limited. Moreover, the unavailability of values in all features also may partially due to unnecessity in performing all the tests as the health of potential patient can be observed from a subset of feature values.

As a concrete example, a dataset we use for evaluation contains clinical diagnostic laboratory test (e.g., routine blood test) results of potential Brazilian COVID-19 patients. Each record corresponding to a potential patient contains 106 features (i.e., 37 categorical and 69 numerical features). It is not unusual to have some but not all values for these 106 features. In other words, it is unnecessary for a potential patient to conduct *all* clinical diagnostic laboratory tests in most cases. More often, only a few essential tests are conducted. This results in some but not all values for the 106 features.

Hence, from the practical viewpoint, it can be expensive and time-consuming to obtain values to most (if not all) of these attributes. For COVID-19 records, it may imply many lab tests, some of which can be invasive and take days to obtain the results. Note that COVID-19 patients can be asymptomatic. Not being able to detect and classify these asymptomatic COVID-19 patients may pose dangers to the society. Hence, it is desirable to be able to identify and classify patients into positive and negative cases *efficiently* and *accurately*. One way to achieve such a goal is to be able to predict the class label for COVID-19 cases, and to be able to do so quickly.

We observe that values for many features are unavailable in some real-life COVID-19 data, and the distribution of these missing values is not uniform. In other words, values for certain features may be unavailable in some records and values for other features become unavailable in some other records. This motivates our third prediction model. Specifically, instead of training the prediction model based on samples drawn from the entire set D of the limited data, our third model—say, $simHA(P_l)$—first divides the original data D of limited data into several disjoint clusters $P_1, \ldots, P_l$ (such that $D = P_1 \cup \ldots \cup P_l$) based on the features with missing values. Then, it trains the autoencoder and the few labeled data samples simultaneously for each cluster P_l. By doing so, from the computational prospective, this model makes predictions based on data in each cluster P_l, which is more focused in terms of common or similarity in feature values. This may imply few valued-features. From the practical healthcare and medical science prospective, this model can be trained on few valued-features, which leads to fewer lab tests are needed to obtain values for those features. Hence, it helps reduce the workload for medical service providers by avoiding to conduct too many lab tests. As such, the patients do not need to conduct many lab tests, but just a small subset of lab tests that are suffice to obtain values for the subset of attributes. As a preview, among 106 features in the Brazilian COVID-19 dataset used for the evaluation in Sect. 4, our system uses about 5, 20 and 40 valued attributes to train our prediction model. The saved time and resources can be used to test more potential COVID-19 patients.

4 Evaluation

To evaluate our health analytics system with three prediction models, we conducted experiments on a real-life open COVID-19 data[2]. Specifically, this dataset contains privacy-preserving samples collected from results of the SARS-CoV-2 reverse transcription polymerase chain reaction (RT-PCR) and additional clinic diagnostic laboratory tests (e.g., routine blood and urine tests) during a hospital visit in the state of São Paulo, Brazil.

We first preprocessed the dataset by performing a standard scaler normalization. The resulting dataset contains 5,644 potential patients (COVID-19 or

---

[2] https://www.kaggle.com/einsteindata4u/covid19.

74   C. K. Leung et al.

Table 1. Evaluation results (F1 scores) on prediction made based on samples from the entire set D of limited data (with the highest F1 score of each column highlighted in **bold**).

#samples per class	1	5
Linear SVC [63]	0.52	0.53
Random forest [63]	0.52	0.53
Logistic regression [63]	0.52	0.56
Gradient boost [63]	0.53	0.62
Our $serHA(D)$	0.62	0.63
Our $simHA(D)$	**0.90**	**0.90**

not) or samples or instances. Each instance consists of a privacy-preserving randomized sample ID, age group, hospitalization status (e.g., admitted to intensive care unit (ICU), semi-intensive unit (SIU), or regular ward), 106 other features (including 37 categorical and 69 numerical features), and a class label (i.e., tested COVID-19 positive or not), for a total of 112 attributes. In term of data distribution, 558 instances (i.e., 9.9% of 5,644 instances in the dataset) were tested/labelled "positive" for the SARS-CoV-2 test result and the remaining 5,086 instances (i.e., 90.1%) were tested/labelled "negative".

Afterwards, we tested our prediction models on different number of labeled samples {1, 5} per class label. With two class labels (i.e., COVID-19 positive and COVID-19 negative), we conducted our experiments with {2, 10} samples from the dataset. We set 10 data samples for each class as the test samples. We also applied first-order gradient-based optimization of stochastic objective functions—called ADAptive Moment estimation (Adam) optimizer [62]. For experiments, the learning rate and batch size were set to 0.001 and 32, respectively.

Moreover, we also compared our system with related ML algorithms including linear support vector classification (SVC), random forest, logistic regression, and gradient boosting [63].

Table 1 shows evaluation results of our two prediction models—namely, $serHA(D)$ and $simHA(D)$, which both made predictions based on samples from the entire set D of limited data—when compared with related ML algorithms. The results reveal that, when using 1 or 5 samples per class (i.e., 2 or 10 samples from D), our $serHA(D)$ led to higher F1 score than the ML algorithms. By simultaneously training both input feature reconstruction and COVID-19 outcome prediction, our $simHA(D)$ led to the highest F1 score. Note that F1 score is computed as:

$$F1 = \frac{2 \text{ prec} \times \text{recall}}{\text{prec} + \text{recall}} = \frac{2TP}{2TP + FP + FN} \tag{5}$$

where TP is true positives, FP is false positives, and FN is false negatives. A maximum possible F1 score of 1 indicates an accurate prediction with perfect precision and recall. Precision measures the fraction of TP among those predicted

Fig. 3. Our third prediction model divides the data into four clusters.

Table 2. Evaluation results (F1 scores) on prediction made based on samples from focused partition P_1 of limited data.

#samples per class	1	5
Linear SVC [63]	0.46	0.58
Random forest [63]	0.52	0.85
Logistic regression [63]	0.60	0.82
Gradient boost [63]	0.51	0.59
Our $simHA(P_1)$	**0.93**	**0.94**

to be positive, i.e., prec $= \frac{TP}{TP+FP}$. Recall measures the fraction of TP among those actual positive cases, i.e., recall $= \frac{TP}{TP+FN}$.

With high F1 score from $simHA(D)$, we split the entire set D of limited data of 5,644 potential COVID-19 patients into 4 disjoint partitions (i.e., $D = P_0 \cup P_1 \cup P_2 \cup P_3$) by k-means clustering. See Fig. 3. Then, we train $simHA(P_l)$ based on samples drawn from P_l. For example, P_0 contains records with 5 valued attributes, P_1 contains records with 20 valued attributes (e.g., type-A and type-B flu viruses), P_2 contains records with other 20 valued attributes (e.g., hemoglobin, platelets), and P_3 contains records with 40 valued attributes (e.g., respiratory syncytial virus (RSV)).

Table 2 shows evaluation results of our third prediction model—namely, $simHA(P_1)$, which made based on samples from a partition P_1 of the entire set D of limited data. The results reveal that, when using 1 or 5 samples per class (i.e., 2 or 10 samples from P_1), by simultaneous training based the focused on focused partition P_1 of limited data, our $simHA(P_1)$ led to the highest F1 score. Similar results obtained for other three partitions.

5   Conclusions

In this paper, we presented a data science system—consisting of three prediction models—for health analytics on COVID-19 data with few-shot learning. The first (serial) model $serHA(D)$ trains the autoencoder for reconstructing input features before training the prediction module for making COVID-19 outcome prediction (i.e., COVID-19 positive or not), by using a few (e.g., 1 or 5) samples drawn from a set D of limited data. The second model $simHA(D)$ simultaneously trains the autoencoder and prediction module. The third model $simHA(P_l)$ uses samples drawn from focused partitions P_l of D. Evaluation on real-life COVID-19 open data capturing clinical diagnostic laboratory test (e.g., routine blood test) results reveals that all three models led to higher F1 score than related machine learning algorithms. Among the three, $simHA(P_l)$ led to the highest F1 score. This demonstrates the usefulness and practicality of our system in health analytics on COVID-19 data. As *ongoing and future work*, we explore ways to further enhance the prediction accuracy. We also would transfer our learned knowledge to health analytics on other healthcare and disease data.

Acknowledgement. This work is partially supported by NSERC (Canada) and University of Manitoba.

References

1. Korfkamp, D., Gudenkauf, S., Rohde, M., Sirri, E., Kieschke, J., Appelrath, H.-J.: Opening up data analysis for medical health services: cancer survival analysis with CARESS. In: Bellatreche, L., Mohania, M.K. (eds.) DaWaK 2014. LNCS, vol. 8646, pp. 382–393. Springer, Cham (2014). https://doi.org/10.1007/978-3-319-10160-6_34
2. Leung, C.K., et al.: Explainable data analytics for disease and healthcare informatics. In: IDEAS 2021, pp. 12:1–12:10 (2021)
3. Olawoyin, A.M., Leung, C.K., Choudhury, R.: Privacy-preserving spatio-temporal patient data publishing. In: Hartmann, S., Küng, J., Kotsis, G., Tjoa, A.M., Khalil, I. (eds.) DEXA 2020, Part II. LNCS, vol. 12392, pp. 407–416. Springer, Cham (2020). https://doi.org/10.1007/978-3-030-59051-2_28
4. Shang, S., et al.: Spatial data science of COVID-19 data. In: IEEE HPCC-SmartCity-DSS 2020, pp. 1370–1375 (2020)
5. Jiang, F., Leung, C.K.-S.: Stream mining of frequent patterns from delayed batches of uncertain data. In: Bellatreche, L., Mohania, M.K. (eds.) DaWaK 2013. LNCS, vol. 8057, pp. 209–221. Springer, Heidelberg (2013). https://doi.org/10.1007/978-3-642-40131-2_18
6. Lakshmanan, L.V.S., Sadri, F.: Modeling uncertainty in deductive databases. In: Karagiannis, D. (ed.) DEXA 1994. LNCS, vol. 856, pp. 724–733. Springer, Heidelberg (1994). https://doi.org/10.1007/3-540-58435-8_238
7. Leung, C.K.-S., MacKinnon, R.K.: Balancing tree size and accuracy in fast mining of uncertain frequent patterns. In: Madria, S., Hara, T. (eds.) DaWaK 2015. LNCS, vol. 9263, pp. 57–69. Springer, Cham (2015). https://doi.org/10.1007/978-3-319-22729-0_5
8. Leung, C.K.-S., MacKinnon, R.K.: BLIMP: a compact tree structure for uncertain frequent pattern mining. In: Bellatreche, L., Mohania, M.K. (eds.) DaWaK 2014. LNCS, vol. 8646, pp. 115–123. Springer, Cham (2014). https://doi.org/10.1007/978-3-319-10160-6_11
9. Braun, P., Cuzzocrea, A., Jiang, F., Leung, C.K.-S., Pazdor, A.G.M.: MapReduce-based complex big data analytics over uncertain and imprecise social networks. In: Bellatreche, L., Chakravarthy, S. (eds.) DaWaK 2017. LNCS, vol. 10440, pp. 130–145. Springer, Cham (2017). https://doi.org/10.1007/978-3-319-64283-3_10
10. Jiang, F., et al.: Finding popular friends in social networks. In: CGC 2012, pp. 501–508 (2012)

11. Jiang, F., Leung, C.K.-S.: Mining interesting "following" patterns from social networks. In: Bellatreche, L., Mohania, M.K. (eds.) DaWaK 2014. LNCS, vol. 8646, pp. 308–319. Springer, Cham (2014). https://doi.org/10.1007/978-3-319-10160-6_28

12. Kumar, N., Chandarana, Y., Anand, K., Singh, M.: Using social media for word-of-mouth marketing. In: Bellatreche, L., Chakravarthy, S. (eds.) DaWaK 2017. LNCS, vol. 10440, pp. 391–406. Springer, Cham (2017). https://doi.org/10.1007/978-3-319-64283-3_29

13. Leung, C.K., et al.: Parallel social network mining for interesting 'following' patterns. CCPE **28**(15), 3994–4012 (2016)

14. Leung, C.K.-S., Jiang, F.: Big data analytics of social networks for the discovery of "following" patterns. In: Madria, S., Hara, T. (eds.) DaWaK 2015. LNCS, vol. 9263, pp. 123–135. Springer, Cham (2015). https://doi.org/10.1007/978-3-319-22729-0_10

15. Singh, S.P., et al.: A theoretical approach to discover mutual friendships from social graph networks. In: iiWAS 2019, pp. 212–221 (2019)

16. Chanda, A.K., et al.: A new framework for mining weighted periodic patterns in time series databases. ESWA **79**, 207–224 (2017)

17. Cuzzocrea, A., Jiang, F., Leung, C.K., Liu, D., Peddle, A., Tanbeer, S.K.: Mining popular patterns: a novel mining problem and its application to static transactional databases and dynamic data streams. In: Hameurlain, A., Küng, J., Wagner, R., Cuzzocrea, A., Dayal, U. (eds.) Transactions on Large-Scale Data- and Knowledge-Centered Systems XXI. LNCS, vol. 9260, pp. 115–139. Springer, Heidelberg (2015). https://doi.org/10.1007/978-3-662-47804-2_6

18. Ishita, S.Z., et al.: New approaches for mining regular high utility sequential patterns. Appl. Intell. (2021). https://doi.org/10.1007/s10489-021-02536-7

19. Leung, C.K.-S., Jiang, F.: Frequent pattern mining from time-fading streams of uncertain data. In: Cuzzocrea, A., Dayal, U. (eds.) DaWaK 2011. LNCS, vol. 6862, pp. 252–264. Springer, Heidelberg (2011). https://doi.org/10.1007/978-3-642-23544-3_19

20. Jackson, M.D., et al.: A Bayesian framework for supporting predictive analytics over big transportation data. In: IEEE COMPSAC 2021, pp. 332–337 (2021)

21. Leung, C.K., Braun, P., Pazdor, A.G.M.: Effective classification of ground transportation modes for urban data mining in smart cities. In: Ordonez, C., Bellatreche, L. (eds.) DaWaK 2018. LNCS, vol. 11031, pp. 83–97. Springer, Cham (2018). https://doi.org/10.1007/978-3-319-98539-8_7

22. Leung, C.K., Braun, P., Hoi, C.S.H., Souza, J., Cuzzocrea, A.: Urban analytics of big transportation data for supporting smart cities. In: Ordonez, C., Song, I.-Y., Anderst-Kotsis, G., Tjoa, A.M., Khalil, I. (eds.) DaWaK 2019. LNCS, vol. 11708, pp. 24–33. Springer, Cham (2019). https://doi.org/10.1007/978-3-030-27520-4_3

23. Terroso-Saenz, F., Valdes-Vela, M., Skarmeta-Gomez, A.F.: Online urban mobility detection based on velocity features. In: Madria, S., Hara, T. (eds.) DaWaK 2015. LNCS, vol. 9263, pp. 351–362. Springer, Cham (2015). https://doi.org/10.1007/978-3-319-22729-0_27

24. Chen, Y., et al.: A data science solution for supporting social and economic analysis. In: IEEE COMPSAC 2021, pp. 1690–1695 (2021)

25. Leung, C.K., Jiang, F.: A data science solution for mining interesting patterns from uncertain big data. In: IEEE BDCloud 2014, pp. 235–242 (2014)

26. Ordonez, C., Song, I.: Guest editorial - DaWaK 2019 special issue - evolving big data analytics towards data science. DKE **129**, 101838:1–101838:2 (2020)

27. Jiang, F., Leung, C.K.: A data analytic algorithm for managing, querying, and processing uncertain big data in cloud environments. Algorithms 8(4), 1175–1194 (2015)

28. Koohang, A., Nord, J.H.: Critical components of data analytics in organizations: a research model. ESWA **166**, 114118:1–114118:9 (2021)

29. Leung, C.K.: Big data analysis and mining. In: Encyclopedia of Information Science and Technology, 4th edn., pp. 338–348 (2018)

30. Leung, C.K., Zhang, H., Souza, J., Lee, W.: Scalable vertical mining for big data analytics of frequent itemsets. In: Hartmann, S., Ma, H., Hameurlain, A., Pernul, G., Wagner, R.R. (eds.) DEXA 2018, Part I. LNCS, vol. 11029, pp. 3–17. Springer, Cham (2018). https://doi.org/10.1007/978-3-319-98809-2_1

31. Wang, T., et al.: Distributed big data computing for supporting predictive analytics of service requests. In: IEEE COMPSAC 2021, pp. 1724–1729 (2021)

32. Alam, M.T., Ahmed, C.F., Samiullah, M., Leung, C.K.: Discriminating frequent pattern based supervised graph embedding for classification. In: Karlapalem, K., et al. (eds.) PAKDD 2021, Part II. LNCS (LNAI), vol. 12713, pp. 16–28. Springer, Cham (2021). https://doi.org/10.1007/978-3-030-75765-6_2

33. Alam, M.T., Ahmed, C.F., Samiullah, M., Leung, C.K.: Mining frequent patterns from hypergraph databases. In: Karlapalem, K., et al. (eds.) PAKDD 2021, Part II. LNCS (LNAI), vol. 12713, pp. 3–15. Springer, Cham (2021). https://doi.org/10.1007/978-3-030-75765-6_1

34. Fariha, A., Ahmed, C.F., Leung, C.K.-S., Abdullah, S.M., Cao, L.: Mining frequent patterns from human interactions in meetings using directed acyclic graphs. In: Pei, J., Tseng, V.S., Cao, L., Motoda, H., Xu, G. (eds.) PAKDD 2013, Part I. LNCS (LNAI), vol. 7818, pp. 38–49. Springer, Heidelberg (2013). https://doi.org/10.1007/978-3-642-37453-1_4

35. Lee, W., Leung, C.K., Nasridinov, A. (eds.): Big Data Analyses, Services, and Smart Data. AISC, vol. 899. Springer, Singapore (2021). https://doi.org/10.1007/978-981-15-8731-3

36. Leung, C.K.-S.: Uncertain frequent pattern mining. In: Aggarwal, C.C., Han, J. (eds.) Frequent Pattern Mining, pp. 339–367. Springer, Cham (2014). https://doi.org/10.1007/978-3-319-07821-2_14

37. Leung, C.K.-S., Tanbeer, S.K.: Mining popular patterns from transactional databases. In: Cuzzocrea, A., Dayal, U. (eds.) DaWaK 2012. LNCS, vol. 7448, pp. 291–302. Springer, Heidelberg (2012). https://doi.org/10.1007/978-3-642-32584-7_24

38. Roy, K.K., Moon, M.H.H., Rahman, M.M., Ahmed, C.F., Leung, C.K.: Mining sequential patterns in uncertain databases using hierarchical index structure. In: Karlapalem, K., et al. (eds.) PAKDD 2021, Part II. LNCS (LNAI), vol. 12713, pp. 29–41. Springer, Cham (2021). https://doi.org/10.1007/978-3-030-75765-6_3

39. Gan, W., Lin, J.C.-W., Fournier-Viger, P., Chao, H.-C.: Mining recent high-utility patterns from temporal databases with time-sensitive constraint. In: Madria, S., Hara, T. (eds.) DaWaK 2016. LNCS, vol. 9829, pp. 3–18. Springer, Cham (2016). https://doi.org/10.1007/978-3-319-43946-4_1

40. Leung, C.K.: Frequent itemset mining with constraints. In: Encyclopedia of Database Systems, 2nd edn., pp. 1531–1536 (2018)

41. Ahn, S., et al.: A fuzzy logic based machine learning tool for supporting big data business analytics in complex artificial intelligence environments. In: FUZZ-IEEE 2019, pp. 1259–1264 (2019)

42. Al-Amin, S.T., Ordonez, C.: Scalable machine learning on popular analytic languages with parallel data summarization. In: Song, M., Song, I.-Y., Kotsis, G., Tjoa, A.M., Khalil, I. (eds.) DaWaK 2020. LNCS, vol. 12393, pp. 269–284. Springer, Cham (2020). https://doi.org/10.1007/978-3-030-59065-9_22

43. Leung, C.K., et al.: Machine learning and OLAP on big COVID-19 data. In: IEEE BigData 2020, pp. 5118–5127 (2020)

44. Diamantini, C., Potena, D., Storti, E.: Exploiting mathematical structures of statistical measures for comparison of RDF data cubes. In: Bellatreche, L., Chakravarthy, S. (eds.) DaWaK 2017. LNCS, vol. 10440, pp. 33–41. Springer, Cham (2017). https://doi.org/10.1007/978-3-319-64283-3_3

45. Leung, C.K.: Mathematical model for propagation of influence in a social network. In: Encyclopedia of Social Network Analysis and Mining, 2nd edn., pp. 1261–1269 (2018)

46. Kaski, S., Sinkkonen, J., Peltonen, J.: Data visualization and analysis with self-organizing maps in learning metrics. In: Kambayashi, Y., Winiwarter, W., Arikawa, M. (eds.) DaWaK 2001. LNCS, vol. 2114, pp. 162–173. Springer, Heidelberg (2001). https://doi.org/10.1007/3-540-44801-2_17

47. Leung, C.K., Carmichael, C.L.: FpVAT: a visual analytic tool for supporting frequent pattern mining. ACM SIGKDD Explor. **11**(2), 39–48 (2009)

48. Leung, C.K., et al.: A visual data science solution for visualization and visual analytics of big sequential data. In: IV 2021, pp. 224–229 (2021)

49. Arora, N.R., Lee, W., Leung, C.K.-S., Kim, J., Kumar, H.: Efficient fuzzy ranking for keyword search on graphs. In: Liddle, S.W., Schewe, K.-D., Tjoa, A.M., Zhou, X. (eds.) DEXA 2012, Part I. LNCS, vol. 7446, pp. 502–510. Springer, Heidelberg (2012). https://doi.org/10.1007/978-3-642-32600-4_38

50. Wang, Y., et al.: Generalizing from a few examples: a survey on few-shot learning. ACM CSUR **53**(3), 63:1–63:34 (2020)

51. Snell, J., et al.: Prototypical networks for few-shot learning. In: NIPS 2017, pp. 4077–4087 (2017)

52. Baldi, P.: Autoencoders, unsupervised learning, and deep architectures. PMLR **27**, 37–50 (2012)

53. Wei, R., Mahmood, A.: Recent advances in variational autoencoders with representation learning for biomedical informatics: a survey. IEEE Access **9**, 4939–4956 (2021)

54. Ji, C., et al.: AEMDA: inferring miRNA-disease associations based on deep autoencoder. Bioinformatics **37**(1), 66–72 (2021)

55. Shi, Z., et al.: A representation learning model based on variational inference and graph autoencoder for predicting lncRNA-disease associations. BMC Bioinform. **22**(1), 136:1–136:20 (2021)

56. Gunduz, H.: An efficient dimensionality reduction method using filter-based feature selection and variational autoencoders on Parkinson's disease classification. Biomed. Sig. Process. Control **66**, 102452:1–102452:9 (2021)

57. Rustam, F., et al.: COVID-19 future forecasting using supervised machine learning models. IEEE Access **8**, 101489–101499 (2020)

58. Brinati, D., et al.: Detection of COVID-19 infection from routine blood exams with machine learning: a feasibility study. J. Med. Syst. **44**(8), 135:1–135:12 (2020). https://doi.org/10.1007/s10916-020-01597-4

59. Xu, M., et al.: Accurately differentiating between patients with COVID-19, patients with other viral infections, and healthy individuals: multimodal late fusion learning approach. J. Med. Internet Res. **23**(1), e25535:1–e25535:17 (2021)

60. Caruana, R.: Multitask learning. Mach. Learn. **28**(1), 41–75 (1997). https://doi.org/10.1023/A:1007379606734
61. Zhang, K., Wu, L., Zhu, Z., Deng, J.: A multitask learning model for traffic flow and speed forecasting. IEEE Access **8**, 80707–80715 (2020)
62. Kingma, D.P., Ba, J.: Adam: a method for stochastic optimization. In: ICLR 2015 (2015). arXiv:1412.6980
63. Sammut, C., Webb, G.I.: Encyclopedia of Machine Learning and Data Mining. Springer, New York (2017)

Cognitive Visual Commonsense Reasoning Using Dynamic Working Memory

Xuejiao Tang[1], Xin Huang[2], Wenbin Zhang[2(✉)],
Travers B. Child[3], Qiong Hu[4], Zhen Liu[5], and Ji Zhang[6]

[1] Leibniz University of Hannover, Hanover, Germany
xuejiao.tang@stud.uni-hannover.de
[2] University of Maryland, Baltimore County, Baltimore, USA
{xinh1,wenbinzhang}@umbc.edu
[3] China Europe International Business School, Shanghai, China
t.b.child@ceibs.edu
[4] Auburn University, Auburn, USA
qzh0011@auburn.edu
[5] Guangdong Pharmaceutical University, Guangzhou, China
liu.zhen@gdpu.edu.cn
[6] University of Southern Queensland, Toowoomba, Australia
ji.zhang@usq.edu.au

Abstract. Visual Commonsense Reasoning (VCR) predicts an answer with corresponding rationale, given a question-image input. VCR is a recently introduced visual scene understanding task with a wide range of applications, including visual question answering, automated vehicle systems, and clinical decision support. Previous approaches to solving the VCR task generally rely on pre-training or exploiting memory with long dependency relationship encoded models. However, these approaches suffer from a lack of generalizability and prior knowledge. In this paper we propose a dynamic working memory based cognitive VCR network, which stores accumulated commonsense between sentences to provide prior knowledge for inference. Extensive experiments show that the proposed model yields significant improvements over existing methods on the benchmark VCR dataset. Moreover, the proposed model provides intuitive interpretation into visual commonsense reasoning. A Python implementation of our mechanism is publicly available at https://github.com/tanjatang/DMVCR

1 Introduction

Reflecting the success of Question Answering (QA) [1] research in Natural Language Processing (NLP), many practical applications have appeared in daily life, such as Artificial Intelligence (AI) customer support, Siri, Alex, etc. However, the ideal AI application is a multimodal system integrating information from different sources [2]. For example, search engines may require more than just text, with image inputs also necessary to yield more comprehensive results. In this respect, researchers have begun to focus on multimodal learning which bridges vision and language processing. Multimodal learning has gained broad interest from the computer vision and natural language processing communities, resulting in the study of Visual Question Answering (VQA) [3]. VQA systems predict answers to language questions conditioned on an image or video. This is challenging for the visual system as often the answer does not directly refer to the image or video in question. Accordingly, high demand has arisen for AI models with cognition-level scene understanding of the real world. But presently, cognition-level scene understanding remains an open, challenging problem. To tackle this problem, Rowan Zeller et al. [4] developed Visual Commonsense Reasoning (VCR). Given an image, a list of object regions, and a question, a VCR model answers the question and provides a rationale for its answer (both the answer and rationale are selected from a set of four candidates). As such, VCR needs not only to tackle the VQA task (i.e., to predict answers based on a given image and question), but also provides explanations for why the given answer is correct. VCR thus expands the VQA task, thereby improving cognition-level scene understanding. Effectively, the VCR task is more challenging as it requires high-level inference ability to predict rationales for a given scenario (i.e., it must infer deep-level meaning behind a scene).

The VCR task is challenging as it requires higher-order cognition and commonsense reasoning ability about the real world. For instance, looking at an image, the model needs to identify the objects of interest and potentially infer people's actions, mental states, professions, or intentions. This task can be relatively easy for human beings in most situations, but it remains challenging for up-to-date AI systems. Recently, many researchers have studied VCR tasks (see, e.g., [4–8]). However, existing methods focus on designing reasoning modules without consideration of prior knowledge or pre-training the model on large scale datasets which lacks generalizability. To address the aforementioned challenges, we propose a Dynamic working Memory based cognitive Visual Commonsense Reasoning network (DMVCR), which aims to design a network mimicking human thinking by storing learned knowledge in a dictionary (with the dictionary regarded as prior knowledge for the network). In summary, our main contributions are as follows. First, we propose a new framework for VCR. Second, we design a dynamic working memory module with enhanced run-time inference for reasoning tasks. And third, we conduct a detailed experimental evaluation on the VCR dataset, demonstrating the effectiveness of our proposed DMVCR model.

The rest of this paper is organized as follows. In Sect. 2 we review related work on QA (and specifically on VCR). Section 3 briefly covers notation. In Sect. 4 we detail how the VCR task is tackled with a dictionary, and how we train a dictionary to assist inference for reasoning. In Sect. 5 we apply our model to the VCR dataset. Finally, in Sect. 6 we conclude our paper.

2 Related Work

Question answering (QA) has become an increasingly important research theme in recent publications. Due to its broad range of applications in customer service and smart question answering, researchers have devised several QA tasks (e.g., Visual Question Answering (VQA) [3], Question-Answer-Generation [9]). Recently, a new QA task named VCR [4] provides answers with justifications for questions accompanied by an image. The key step in solving the VCR task is to achieve inference ability. There exists two major methods of enhancing inference ability. The first focuses on encoding the relationship between sentences using sequence-to-sequence based encoding methods. These methods infer rationales by encoding the long dependency relationship between sentences (see, e.g., R2C [4] and TAB-VCR [6]). However, these models face difficulty reasoning with prior knowledge, and it is hard for them to infer reason based on commonsense about the world. The second method focuses on pre-training [7,8,10]. Such studies typically leverage pre-training models on more than three other image-text datasets to learn various abilities like masked multimodal modeling, and multimodal alignment prediction [8]. The approach then regards VCR as a downstream fine-tuning task. This method however lacks generalizability.

Considering the disadvantages of either aforementioned approach, we design a network which provides prior knowledge to enhance inference ability for reasoning. The idea is borrowed from human beings' experience – prior knowledge or commonsense provides rationale information when people infer a scene. To achieve this goal, we propose a working memory based dictionary module for run-time inference. Recent works such as [11–13] have successfully applied the working memory into QA, VQA, and image caption. Working memory provides a dynamic knowledge base in these studies. However, existing work focuses on textual question answering tasks, paying less attention to inference ability [11,12]. Concretely, the DMN network proposed in [14] uses working memory to predict answers based on given textual information. This constitutes a step forward in demonstrating the power of dynamic memory in QA tasks. However, that approach can only tackle textual QA tasks. Another work in [12] improves upon DMN by adding an input fusion layer (containing textual and visual information) to be used in VQA tasks. However, both methods failed to prove the inference ability of dynamic working memory. Our paper proposes a dictionary unit based on dynamic working memory to store commonsense as prior knowledge for inference.

3 Notations and Problem Formulation

The VCR dataset consists of millions of labeled subsets. Each subset is composed of an image with one to three associated questions. Each question is then associated with four candidate answers and four candidate rationales. The overarching task is formulated as three subtasks: (1) predicting the correct answer for a given question and image ($Q \rightarrow A$); (2) predicting the correct rationale for a given question, image, and correct answer $(QA \rightarrow R)$; and (3) predicting the correct answer and rationale for a given image and question $(Q \rightarrow AR)$. Additionally, we defined two language inputs - query $q\{q_1, q_2, \cdots, q_n\}$ and response $r\{r_1, r_2, \cdots, r_n\}$, as reflected in Fig. 1. In the $Q \rightarrow A$ subtask, query q is the question and response r is the answers. In the $QA \rightarrow R$ subtask, query q becomes the question together with correct answer, while rationales constitute the response r.

4   Proposed Framework

As shown in Fig. 1, our framework consists of four layers: a feature representation layer, a multimodal fusion layer, an encoder layer, and a prediction layer. The first layer captures language and image features, and converts them into dense representations. The represented features are then fed into the multimodal fusion layer to generate meaningful contexts of language-image fused information. Next, the fused features are fed into an encoder layer, which consists of a long dependency encoder [15] RNN module along with a dictionary unit. Finally, a prediction layer is designed to predict the correct answer or rationale.

4.1   Feature Representation Layer

The feature representation layer converts features from images and language into dense representations. For the language, we learn embeddings for the query $q\{q_1, q_2, \cdots, q_n\}$ and response $r\{r_1, r_2, \cdots, r_n\}$ features. Additionally, the object features $o\{o_1, o_2, \cdots, o_n\}$ are extracted from a deep network based on residual learning [16].

Language Embedding. The language embeddings are obtained by transforming raw input sentences into low-dimensional embeddings. The query represented by $q\{q_1, q_2, \cdots, q_n\}$ refers to a question in the question answering task $(Q \rightarrow A)$, and a question paired with correct answer in the reasononing task $(QA \rightarrow R)$. Responses $r\{r_1, r_2, \cdots, r_n\}$ refer to answer candidates in the question answering task $(Q \rightarrow A)$, and rationale candidates in the reasoning task $(QA \rightarrow R)$. The embeddings are extracted using an attention mechanism with parallel structure [17]. Note that the sentences contain tags related to objects in the image. For example, see Fig. 3(a) and the question "Are [0, 1] happy to be here?" The [0, 1] are tags set to identify objects in the image (i.e., the object features of person 1 and person 2).

Object Embedding. The images are filtered from movie clips. To ensure images with rich information, a filter is set to select images with more than two objects each [4]. The object features are then extracted with a residual connected deep network [16]. The output of the deep network is object features with low-dimensional embeddings $o\{o_1, o_2, \cdots, o_n\}$.

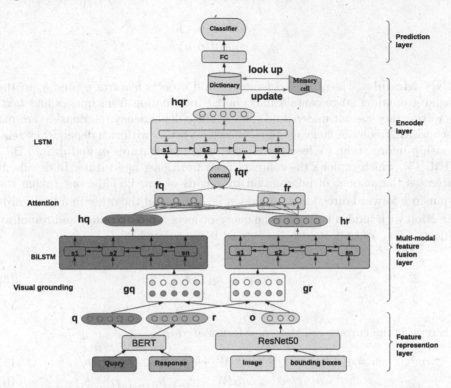

Fig. 1. High-level overview of the proposed DMVCR consisting of four modules: feature representation layer to extract visual and textual features; multimodal feature fusion to contextualize multimodal representations; encoder layer to encode rich visual commonsense; and prediction layer to select the most related response.

4.2 Multimodal Feature Fusion Layer

The multimodal feature fusion layer consists of three modules: a visual grounding module, an RNN module, and an attention module.

Visual Grounding. Visual grounding aims at finding out target objects for query and response in images. As mentioned in Sect. 4.1, tags are set in query and responses to reference corresponding objects. The object features will be extracted and concatenated to language features at the visual grounding unit to obtain the representations with both image and language information. As shown in Fig. 1, the inputs of visual grounding consist of language ($q\{q_1, q_2, \cdots, q_n\}$ and $r\{r_1, r_2, \cdots, r_n\}$) along with related objects features ($o\{o_1, o_2, \cdots, o_m\}$). The output contains aligned language and objects features (g_q and g_r). The white unit at visual grounding is grounded representations, which contains image and text information. It can be formulated as follows (where *concat* represents the concatenate operation):

$$g_r = concat(o, r) \tag{1}$$
$$g_q = concat(o, q) \tag{2}$$

RNN Module. The grounded language and objects features g_q and g_r at the visual grounding stage contain multimodal information from images and text. However, they cannot understand the semantic dependency relationship around each word. To obtain language-objects mixed vectors with rich dependency relationship information, we feed the aligned language features g_q and g_r into BiL-STM [15], which exploits the contexts from both past and future. In details, it increases the amount of information by means of two LSTMs, one taking the input in a forward direction with hidden layer $\overrightarrow{h_{lt}}$, and the other in a backwards direction with hidden layer $\overleftarrow{h_{lt}}$. The query-objects representations and response-objects h_l $_{l \in \{q,r\}}$ output at each time step t is formulated as:

$$h_{lt} = \overrightarrow{h_{lt}} \oplus \overleftarrow{h_{lt}} \tag{3}$$
$$\overrightarrow{h_{lt}} = o_t \odot \tanh(c_t) \tag{4}$$

where c_t is the current cell state and formulated as:

$$c_t = f_t \odot c_{t-1} + i_t \odot tanh(W_c \cdot [c_{t-1}, h_{l(t-1)}, x_t] + b_c) \tag{5}$$
$$i_t = \sigma(W_i \cdot [c_{t-1}, h_{l(t-1)}, x_t] + b_i) \tag{6}$$
$$o_t = \sigma(W_o \cdot [c_{t-1}, h_{l(t-1)}, x_t] + b_o) \tag{7}$$
$$f_t = \sigma(W_f \cdot [c_{t-1}, h_{l(t-1)}, x_t] + b_f) \tag{8}$$

where i, o, f represent input gate, output gate, and forget gate, respectively, and x_t is the tth input of a sequence. In addition, $W_i, W_o, W_f, W_c, b_c, b_i, b_o, b_f$ are trainable parameters with σ representing the sigmoid activation function [18].

Attention Module. Despite the good learning of BiLSTM in modeling sequential transition patterns, it is unable to fully capture all information from images and languages. Therefore, an attention module is introduced to enhance the RNN module, picking up object features which are ignored in the visual grounding and RNN modules. The attention mechanism on object features $o\{o_1, o_2, \cdots, o_n\}$ and response-objects representations h_r is formulated as:

$$\alpha_{i,j} = softmax(o_i W_r h_{rj}) \tag{9}$$
$$\hat{fr_i} = \sum_j \alpha_{i,j} h_{rj} \tag{10}$$

where i and j represent the position in a sentence, W_r is trainable weight matrix. In addition, this attention step also contextualizes the text through object information.

Furthermore, another attention module is implemented between query-objects representations h_q and response-objects representations h_r, so that the output fused query-objects representation contains weighted information of response-objects representations. It can be formulated as:

$$\alpha_{i,j} = softmax(h_{ri}W_q h_{qj}) \tag{11}$$

$$\hat{fq}_i = \sum_j \alpha_{i,j}h_{qj} \tag{12}$$

where W_q is the trainable weight matrix, i and j denote positions in a sentence.

4.3   Encoder Layer

The encoder layer aims to capture the commonsense between sentences and use it to enhance inference. It is composed of an RNN module and a dictionary module.

RNN Module. An RNN unit encodes the fused queries and responses by long dependency memory cells [19], so that relationships between sentences can be captured. The input is fused query ($\hat{f}_q$) and response ($\hat{f}_r$) features. To encode the relationship between sentences, we concatenate $\hat{f}_q$ and $\hat{f}_r$ at sentence length dimensions as the input of LSTM. Its last output hidden layer contains rich information about commonsense between sentences. At time step t, the outputting representations can be formulated as:

$$h_t = o_t \odot tanh(c_t) \tag{13}$$

where the c_t is formulated the same as in Eq. (5). The difference is that $x_t = concat(\hat{f}_q, \hat{f}_q)$, where concat is the concatenate operation. In addition, the outputting representations h_t is the last hidden layer of LSTM, while the outputting in Eq. (3) is every time step of BiLSTM.

Dictionary Module. Despite effective learning of the RNN unit in modeling the relationship between sentences, it is still limited for run-time inference. We therefore propose a dictionary unit to learn dictionary D, and then use it to look up commonsense for inference. The dictionary is a dynamic knowledge base and is being updated during training. We denote the dictionary as a $d \times k$ matrix $D\{d_1, d_2, ..., d_k\}$, where k is the size of dictionary. The given encoded representation h from RNN module will be encoded using the formulations:

$$\hat{h} = \sum_{k=1}^{K} \alpha_k d_k, \alpha = softmax(D^1 h) \tag{14}$$

where α can be viewed as the "key" idea in memory network [13].

4.4 Prediction Layer

The prediction layer generates a probability distribution of responses from the high-dimension context generated in the encoder layer. It consists of a multi-layer perceptron. VCR is a multi-classification task in which one of the four responses is correct. Therefore, multiclass cross-entropy [20] is applied to complete the prediction.

5 Experimental Results

In this section, we conduct extensive experiments to demonstrate the effectiveness of our proposed DMVCR network for solving VCR tasks. We first introduce the datasets, baseline models, and evaluation metrics of our experiments. Then we compare our model with baseline models, and present an analysis of the impact of the different strategies. Finally, we present an intuitive interpretation of the prediction.

5.1 Experimental Settings

Dataset. The VCR dataset [4] is composed of 290k multiple-choice questions in total (including 290k correct answers, 290k correct rationales, and 110k images). The correct answers and rationales labeled in the dataset are met with 90% of human agreements. An adversarial matching approach is adopted to obtain counterfactual choices with minimal bias. Each answer contains 7.5 words on average, and each rationale contains 16 words on average. Each set consists of an image, a question, four available answer choices, and four candidate rationales. The correct answer and rationale are given in the dataset.

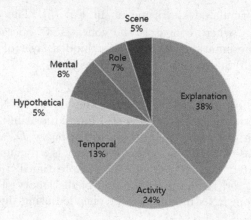

Fig. 2. Overview of the types of inference required by questions in VCR.

The distribution of inference types is shown in Fig. 2. Thirty-eight percent of the inference types are regarding explanation, and 24% of them are about the activity. The rest are related to temporal, mental, role, scene, and hypothetical inference problems.

Hyperparameters. The image features are projected to 512 dimension. The word embedding dimension is 768. The dictionary is a [512,800] matrix, where 512 is the embedding dimension, and 800 is the dictionary size. We separately set the learning rate for the memory cell (the dictionary cell) to 0.02, and others to 0.0002. In addition, for the $Q \rightarrow A$ subtask, we set the hidden size of LSTM encoder to 512. For the $QA \rightarrow R$ subtask, we set the hidden size of LSTM encoder to 64. The model was trained with the Adam algorithm [21] using PyTorch on NVIDIA GPU GTX 1080.

Metric. The VCR task can be regarded as a multi-classification problem. We use mAp [22] to evaluate the performance, which is a common metric for evaluating prediction accuracy in multi-classification areas.

Approach for Comparison. We compare the proposed DMVCR with recent deep learning-based models for VCR. Specifically, the following baseline approaches are evaluated:

- **RevisitedVQA** [23]: Different from the recently proposed systems, which have a reasoning module that includes an attention mechanism or memory mechanism, RevisitedVQA focuses on developing a "simple" alternative model, which reasons the response using logistic regressions and multi-layer perceptrons (MLP).
- **BottomUpTopDown** [24]: Proposed a bottom-up and top-down attention method to determine the feature weightings for prediction. It computes a weighted sum over image locations to fuse image and language information so that the model can predict the answer based on a given scene and question.
- **MLB** [25]: Proposed a low-rank bilinear pooling for the task. The bilinear pooling is realized by using the Hadamard product for attention mechanism and has two linear mappings without biases for embedding input vectors.
- **MUTAN** [26]: Proposed a multimodal fusion module tucker decomposition (a 3-way tensor), to fuse image and language information. In addition, multimodal low-rank bilinear (MLB) is used to reason the response for the input.
- **R2C** [4]: Proposed a fusion module, a contextualization module, and a reasoning module for VCR. It is based on the sequence relationship model LSTM and attention mechanism.

5.2   Analysis of Experimental Results

Task Description. We implement the experiments separately in three steps. We firstly conducted $Q \rightarrow A$ evaluation, and then $QA \rightarrow R$. Finally, we join the $Q \rightarrow A$ result and $QA \rightarrow R$ results to obtain the final $Q \rightarrow AR$ prediction result. The difference between the implementation of $Q \rightarrow A$ and $QA \rightarrow R$ tasks is the input query and response. For the $Q \to A$ task, the query is the paired question, image, four candidate answers; while the response is the correct answer. For the $QA \to R$ task, the query is the paired question, image, correct answer, and four candidate rationales; while the response is the correct rationale.

Table 1. Comparison of results between our methods and other popular methods using the VCR Dataset. The best performance of the compared methods is highlighted. Percentage in parenthesis is our relative improvement over the performance of the best baseline method.

Models	$Q \to A$	$QA \to R$	$Q \to AR$
RevisitedVQA [23]	39.4	34.0	13.5
BottomUpTopDown [24]	42.8	25.1	10.7
MLB [25]	45.5	36.1	17
MUTAN [26]	44.4	32.0	14.6
R2C (Baseline) [4]	**61.9**	**62.8**	**39.1**
DMVCR	**62.4** (+0.8%)	**67.5** (+7.5%)	**42.3** (+8.2%)

Analysis. We evaluated our method on the VCR dataset and compared the performance with other popular models. As the results in Table 1 show, our approach outperforms in all of the subtasks: $Q \to A$, $QA \to R$, and $Q \to AR$. Specifically, our method outperforms MUTAN and MLB by a large margin. Furthermore, it also performs better than R2C.

5.3 Qualitative Results

We evaluate qualitative results on the DMVCR model. The qualitative examples are provided in Fig. 3. The candidate in green represents the correct choice; the candidate with a checkmark ✓ represents the prediction result by our proposed DMVCR model. As the qualitative results show, the DMVCR model improves its power in inference.

For instance, see Fig. 3(a). The question listed is: "What kind of profession does [0, 1] and [2] practice?". The predicted answer is D - "They are all lawyers." Furthermore, the model offers rationale C - "[0, 1] and [2] are all dressed in suits and holding papers or briefcases, and meet with people to discuss their cases." DMVCR correctly infers the rationale based on dress and activity, even though this task is difficult for humans.

DMVCR can also identify human beings' expressions and infer emotion. See for example the result in Fig. 3(b). Question 1 is: "Are [0, 1] happy to be there?". Our model selects the correct answer B along with reason A: "No, neither of them is happy, and they want to go home"; because "[0] looks distressed, not at all happy."

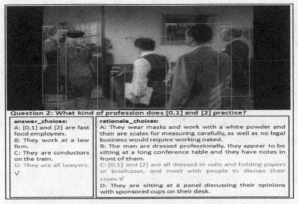

(a) Qualitative example 1. The model predicts the correct answer and rationale.

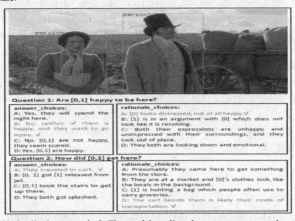

(b) Qualitative example 2. The model predicts the correct answer and rationale.

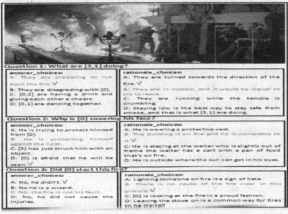

(c) Qualitative example 3. The model predicts the correct answer but incorrect rationale in Question 1. The model predicts an incorrect answer but correct rationale in Question 2.

Fig. 3. Qualitative examples. Prediction from DMVCR is marked with a √ while correct results are highlighted in green. (Color figure online)

Finally, there are also results which predict the correct answer but infer the wrong reason. For instance, see question 1 in Fig. 3(c): "What are [3,1] doing?" DMVCR predicts the correct answer A - "They are preparing to run from the fire." But it infers the wrong reason A - "They are turned towards the direction of the fire." The correct answer is of course B - "They are in motion, and it would be logistical to try to leave." It is also possible for the model to predict a wrong answer but correct rationale. This appears in question 2 of Fig. 3(c). The model predicts the wrong answer D - "[0] is afraid that he will be seen." The correct reason is B - "The building is on fire and he is vulnerable to it."

6   Conclusion

This paper has studied the popular visual commonsense reasoning (VCR). We propose a working memory based model composed of a feature representation layer to capture multiple features containing language and objects information; a multimodal fusion layer to fuse features from language and images; an encoder layer to encode commonsense between sentences and enhance inference ability using dynamic knowledge from a dictionary unit; and a prediction layer to predict a correct response from four choices. We conduct extensive experiments on the VCR dataset to demonstrate the effectiveness of our model and present intuitive interpretation. In the future, it would be interesting to investigate multimodal feature fusion methods as well as encoding commonsense using an attention mechanism to improve the performance of VCR.

References

1. Hirschman, L., Gaizauskas, R.: Natural language question answering: the view from here. Nat. Lang. Eng. **7**(4), 275 (2001)
2. Sharma, P., Ding, N., Goodman, S., Soricut, R.: Conceptual captions: a cleaned, hypernymed, image alt-text dataset for automatic image captioning. In: Proceedings of the 56th Annual Meeting of the Association for Computational Linguistics (Volume 1: Long Papers), pp. 2556–2565 (2018)
3. Antol, S., Agrawal, A., et al.: VQA: visual question answering. In: Proceedings of the IEEE International Conference on Computer Vision, pp. 2425–2433 (2015)
4. Zellers, R., Bisk, Y., et al.: From recognition to cognition: visual commonsense reasoning. In: Proceedings of the IEEE Conference on CVPR (2019)
5. Ben-younes, H., Cadène, R., Cord, M., Thome, N.: MUTAN: multimodal tucker fusion for visual question answering, CoRR (2017)
6. Lin, J., Jain, U., et al.: TAB-VCR: tags and attributes based VCR baselines
7. Yu, F., Tang, J., et al.: ERNIE-ViL: knowledge enhanced vision-language representations through scene graph (2020)
8. Lu, J., Batra, D., et al.: ViLBERT: pretraining task-agnostic visiolinguistic representations for vision-and-language tasks. In: Advances in Neural Information Processing Systems, pp. 13–23 (2019)
9. Lee, D.B., Lee, S., Jeong, W.T., Kim, D., Hwang, S.J.: Generating diverse and consistent QA pairs from contexts with information-maximizing hierarchical conditional VAEs, arXiv preprint arXiv:2005.13837 (2020)

10. Chen, Y.-C., Li, L., et al.: UNITER: Learning universal image-text representations (2019)
11. Sabes, P.N., Jordan, M.I.: Advances in neural information processing systems. In: Tesauro, G., Touretzky, D., Leed, T. (eds.) Advances in Neural Information Processing Systems. Citeseer (1995)
12. Xiong, C., Merity, S., Socher, R.: Dynamic memory networks for visual and textual question answering. In: International Conference on Machine Learning. PMLR, pp. 2397–2406 (2016)
13. Yang, X., Tang, K., et al.: Auto-encoding scene graphs for image captioning. In: Proceedings of the IEEE Conference on CVPR, pp. 10 685–10 694 (2019)
14. Kumar, A., et al.: Ask me anything: dynamic memory networks for natural language processing. In: International Conference on Machine Learning. PMLR, pp. 1378–1387 (2016)
15. Huang, Z., Xu, et al.: Bidirectional LSTM-CRF models for sequence tagging, arXiv preprint arXiv:1508.01991 (2015)
16. He, K., Zhang, X., Ren, S., Sun, J.: Deep residual learning for image recognition. In: Proceedings of the IEEE Conference on CVPR, pp. 770–778 (2016)
17. Devlin, J., Chang, M.-W., et al.: BERT: pre-training of deep bidirectional transformers for language understanding, arXiv preprint arXiv:1810.04805 (2018)
18. Yin, X., Goudriaan, J., Lantinga, E.A., Vos, J., Spiertz, H.J.: A flexible sigmoid function of determinate growth. Ann. Bot. **91**(3), 361–371 (2003)
19. Hochreiter, S., Schmidhuber, J.: Long short-term memory. Neural Comput. **9**, 1735–1780 (1997)
20. Rubinstein, R.: The cross-entropy method for combinatorial and continuous optimization. Methodol. Comput. Appl. Probab. **1**(2), 127–190 (1999)
21. Kingma, D.P., Ba, J.: Adam: a method for stochastic optimization, arXiv preprint arXiv:1412.6980 (2014)
22. Henderson, P., Ferrari, V.: End-to-end training of object class detectors for mean average precision. In: Lai, S.-H., Lepetit, V., Nishino, K., Sato, Y. (eds.) ACCV 2016. LNCS, vol. 10115, pp. 198–213. Springer, Cham (2017). https://doi.org/10.1007/978-3-319-54193-8_13
23. Jabri, A., Joulin, A., Van Der Maaten, L.: Revisiting visual question answering baselines. In: Leibe, B., Matas, J., Sebe, N., Welling, M. (eds.) ECCV 2016. LNCS, vol. 9905, pp. 727–739. Springer, Cham (2016). https://doi.org/10.1007/978-3-319-46448-0
24. Anderson, P., He, X., et al.: Bottom-up and top-down attention for image captioning and visual question answering. In: Proceedings of the IEEE Conference on Computer Vision and Pattern Recognition, pp. 6077–6086 (2018)
25. Kim, J.-H., On, K.-W., Lim, W., Kim, J., Ha, J.-W., Zhang, B.-T.: Hadamard product for low-rank bilinear pooling (2016)
26. Ben-Younes, H., Cadene, R., Cord, M., Thome, N.: MUTAN: multimodal tucker fusion for visual question answering. In: Proceedings of the IEEE International Conference on Computer Vision, pp. 2612–2620 (2017)

Knowledge Representation

Universal Storage Adaption for Distributed RDF-Triple Stores

Ahmed Al-Ghezi$^{(\boxtimes)}$ and Lena Wiese◉

Goethe University Frankfurt, Robert-Mayer-Str. 10,
60629 Frankfurt am Main, Germany
{alghezi,lwiese}@cs.uni-frankfurt.de

Abstract. The publication of machine-readable information has been significantly increasing both in the magnitude and complexity of the embedded relations. The Resource Description Framework (RDF) plays a big role in modelling and linking web data and their relations. Dedicated systems (RDF stores/triple stores) were designed to store and query the RDF data. Due to the size of RDF data, a distributed RDF store may use several federated working nodes to store data in a partitioned manner. After partitioning, some of the data need to be replicated to avoid communication cost. In order to efficiently answer queries, each working node needs to put its share of data into multiple indexes. Those indexes have a data-wide size and consume a considerable amount of storage space. The third storage-consuming structure is the join cache – a special index where the frequent join results are cached.

We present a novel adaption approach to the storage management of a distributed RDF store. The system aims to find optimal data assignments to the different indexes, replications, and join cache within the limited storage space. To achieve this, we present a cost model based on the workload that often contains frequent patterns. The workload is dynamically and continuously analyzed to evaluate predefined rules considering the benefits and costs of all options of assigning data to the storage structures. The objective is to reduce query execution time. Our universal adaption approach outperformed the in comparison to state-of-the-art competitor systems.

Keywords: RDF · Workload-aware · Space-adaption

1 Introduction

The Resource Description Framework (RDF) [10] has been widely used to model the data on the web. Despite its simple triple-based structure (each triple consisting of subject, predicate and object), RDF showed a high ability to model the complex relationships between the web entities and preserve their semantic. It provided the scalability that allowed the RDF data to grow big from the range of billions to the range of trillions of triples [17]. As a result, RDF data experienced a rapid increase both in the size and complexity of the embedded

© Springer Nature Switzerland AG 2021
M. Golfarelli et al. (Eds.): DaWaK 2021, LNCS 12925, pp. 97–108, 2021.
https://doi.org/10.1007/978-3-030-86534-4_8

relationships [6]. That emphasises more challenges on the RDF triple stores in terms of managing and structuring that complex and huge data while still showing acceptable query execution performance. These stores have to have multiple data-wide indexes, cache, and replications (in case of distributed triple stores). The highly important constraint to be considered in managing these structures is the storage space. Thus, many research works tried to optimize their usage by depending on workload analysis (see Sect. 3). However, they focused on the problems of indexes, replication, and cache separately, despite the fact that the three optimization problems are basically modeling an integrated single optimization problem. They share the same constraint (the storage space) and objective function (maximizing the performance).

We propose a universal adaption approach for the storage layer of RDF-triple store. It uses the workload to evaluate the benefits of indexes, replication, and cache and selects the most beneficial items to fill the limited storage space.

The rest of this paper is organized as follows: Sect. 2 provides descriptions of the adaption components. Section 3 reviews the related work. In Sect. 4 we formulate our cost model. In Sect. 5 we describe our method of analysing the workload. In Sect. 6 we derive the benefits of the adaptions components. We provide the universal adaption algorithm in Sect. 7. We practically evaluation the approach in Sect. 8. Finally we state our conclusion in Sect. 9.

2   Background

RDF Indexing. The most important data structure in an RDF store is the index. Systems like RDF-3X [16] and Hexastore [20] decided to build the full set of 6 possible indexes to have the full flexibility in query planning. This strategy fully supports the performance at the high expense of storage space. To deal with this high cost, other RDF stores preferred to choose only a limited number of indexes. These indexes are chosen based on some observations of the workload and the system has to live with it. However, the storage availability and workload trends are variable parameters. Thus fixed prior decisions about the indexes might be very far from optimal. Instead, our adaptable system evaluates the status of the workload and space at runtime and adjusts its indexing layer accordingly.

Replications. Due to the huge size and relation complexity of the RDF data, many RDF stores moved towards distributed systems. Instead of a single node, multiple federated nodes are used to host and query the data. However, the RDF data-set needs to be partitioned and assigned to those working nodes. Due to the complexity of this operation, replications are often needed to decrease the communication cost between the nodes. However, those replications may require a lot of storage space, and RDF stores had to manage this problem by trying to select only limited data for replication with the highest expected benefits. That benefit is derived from the workload and the relative locality of the data. However, all of the related works either assume the existence of some initial workload, or fixed parameters and thresholds which are not clearly connected or calculated from the workload. Moreover, the replication and indexes share the same storage space, and the space optimizing process needs to consider both of them in the same process.

Join Cache. Executing a SPARQL query often requires multiple costly join operations. The size of the join results could be much smaller than the processed triples. Moreover, the real workload typically contains frequent patterns [17]. That makes a join cache very beneficial to the performance. However, it consumes a lot of storage space. Since the cache shares the same storage space of indexes and replication, and shares the same objective of query performance, it should be integrated into the same optimization problem.

Workload Analysis. RDF stores have to manage their needs to indexes and replications with the limited storage space. Historical workload played a vital role in such management [1,17]. The analysis of the workload can be classified into *active* and *passive.* The active analysis is carried out on a collected workload aiming to derive its trends, detect future behaviour, then adapt its structures accordingly. However, such an adaption could highly degrade the system performance if the workload does not contain the expected frequent patterns. To avoid this problem, many systems preferred making fixed decisions about their indexes and replications upon passive analysis to some workload samples to draw average behaviour. The systems have to live with these decisions on any status of workload and storage space.

Universal Adaption. Instead of separate space optimization towards replication, indexes, or cache, we put the three types of structures into a single optimization process. The system chooses the most beneficial options to fill its limited storage space. We define a single cost model for the three types based on workload analysis.

3   Related Work

RDF-3X [16] is one of the first native RDF store. it uses an excessive indexing scheme by implementing all the 6 possible index permutations besides extra 3 aggregate indexes. The six-indexes scheme was also by Hexastore [20]. To decrease the storage overhead, RDF-3X uses a dictionary [4], where each textual element in the RDF data set is mapped to a small integer code. The **H-RDF-3X** system by **Huang et al.** [12] was the first distributed system that used a grid of nodes, such that each node is hosting an RDF-3X triple store. The data is partitioned using METIS graph partitioning [13]. To reduce the communication cost, H-RDF-3X forces uniform k-hop replication on the partitions' borders. Any query shorter than k is guaranteed to be executed locally. Unfortunately, the storage overhead of the replication increases exponentially with k, and H-RDF-3X did not provide any systematic method to practically calculate the optimal value of k. **Partout** [7] implemented workload awareness on the level of partitioning.

It horizontally partitions the data set inspired by the classical approaches of partitioning relational tables [3]. The system tries to assign the most related fragments to same partitions. Unfortunately, the results of such a partitioning are highly affected by the quality of the used workload. It could end up with small fragments representing the workload and a big single fragment containing everything else. **WARP** [11] proposed to use a combination of Partout and H-RDF-3X. Initially, the system is partitioned and replicated using the H-RDF-3X approach with a *small* value of k. Then, it uses workload to recognize the highest accessed triples for further border replications. Besides lacking the methodology to determine k, WARP supposes sharp threshold between frequent and non frequent triples. **Peng** et al. [17] proposed a partitioning approach inspired by Partout [7] but supported by replications. **AdPart** [9] is an in-memory distributed triple store. It aggressively partitions the data set by hashing the subject of each triple. As this is known to produce high communication costs, AdPart proposes two solutions. The first is by updating the dynamic programming algorithm [8,14,16] that is used to find the optimal query execution plan, to include the communication cost. However, this algorithm depends on the accuracy of the cost estimation which is already a challenging issue regarding calculating the optimal join plan in a centralized system like RDF-3X [16]. The second solution to the communication cost problem is by adding workload-driven replications. AdPart collects queries at runtime, builds the workload, and adapts its replications with time dynamically. Yet, AdPart requires a fixed setting of a frequency threshold that is used to differentiate between frequent and non-frequent items, making it a only a semi-automated system. **TriAd** [8] performs hash-based partitioning on both the subject and the object. That allows it to have a two-way locality awareness of each triple. Similar to AdPart, TriAd employs this to decrease the high communication cost caused by the hash partitioning. Aluç et al. [2] uses Tunable-LSH to cluster the workload and assigns the most related patterns to the same page. However, the approach does not count for the distributed replication and join cache.

4   Our Flexible Universal Cost Model

Our system aims to make its storage resources adaptable with the current status of the space and workload. The data set is divided into a set of units called the *consumers*. Each consumer may be assigned to a storage resource equal to its size based on one out of the different index options. The goal now is to find an optimal assignment based on a function that calculates a benefit for each setting. The optimization problem can be reduced to the Knapsack problem [5], where the local storage space is a knapsack of size n, which we want to fill with the most beneficial assignments (the items). However, since that the size of the assignments is small with respect to the total storage size, we may relax the condition of requiring a totally filled knapsack. That allows the problem to be solved greedily (instead of a more costly dynamic programming [15]). That is carried out by dividing the *benefit* of each item by its size and greedily filling the knapsack with most beneficial items. To reduce our model to this problem we need to derive the benefit of each assignment. To achieve the universal adaption, we derive such benefits for indexes, join cache, and replication. However, the effective benefit is related to how often the system is going to use that assignment. We call this the access rate ρ and calculate it from the workload analysis. Based on this, the effective benefit of each assignment g that has a performance benefit η, and an access rate ρ is given by the following:

$$benefit(g) = \frac{\eta(g) \cdot \rho(g)}{size(g)} \tag{1}$$

The size in the above formula is given in number of triples that are affected by the assignment. By applying the formula on a set of assignment options, we may sort them and select the most beneficial option. In the next sections, we describe detecting the access rates and deriving the performance benefits

5   Workload Analysis

The aim of our workload analysis module is to find the access rate for each option of storage assignment – that is, ρ in Formula 1. First, we generate from the workload *general access rules*, which measures the average behaviour (e.g. the average index usage). Second, we detect the frequent items in the workload using a special structure called *heat query* resulting in a set of *specific rules*. The system actively measures the effectiveness of these rules, and prunes the impact of the rule of low effectiveness. By using this approach, our system gets the benefits of frequent patterns in the workload, and avoids their drawbacks by relying on average workload behaviour.

5.1   Heat Queries

A workload is a collection of the previous queries. However, we need to store the workload in a structure that keeps the relationships between the queries as well as their frequencies. A heat query is inspired from the concept of a *heat map* but instead of the matrix of heat values, we have a graph of heat values representing access rates. The workload is then seen as a set of heat queries. The heat query extends the original concept of *global query graph* originally proposed by Partout [7]. Each heat query is implemented as hashed map, where the key is a triple pattern and the value is a statistics object representing the frequency of appearance for that pattern, and the used indexes to evaluate it as well as performance-benefits values explained in the Sect. 6. $heatQueryAccess(v, \chi)$ is a function that returns the heat value of $v \in V$ for the index χ in the heat queries set.

5.2   Heat Query Generation

We explain in this subsection the generation of the heat query set out of a workload Q and an RDF graph G illustrated in Fig. 1. Each time a query q is executed, it forms a new heat query h, with heat (frequency) set to 1 for its vertices. Next, h is either added to the current heat query set H or combined with H, if there is a heat query $h_i \in H$ that has at least one shared element; the shared element is either a single vertex or an entire triple pattern.

When combining two heat queries, the shared vertices become "hotter" by summing up the heats of the heat query graph and the new heat query. The combining process is shown in Fig. 1 for a workload $\langle Q_1, Q_2, Q_3, Q_4 \rangle$. When Q_2 is received, it makes the matching part (to which Q_2 is combined) of the previous heat query hotter – which is illustrated by a darker color. The same applies for Q_3 and Q_4. Any variable in the query (here $?x$, $?y$ and $?z$) is replaced ("unified") by a single variable $?x$ to allow the variables to be directly combined. Note that in the case that a variable subsumes a constant, both the constant and the variable exist as separate vertices in the heat query. This

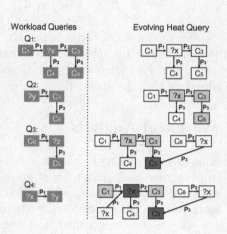

Fig. 1. Heat query evolving from four queries

also happens when Q_4 is combined. It increases the heat value of C_1 in Fig. 1 and creates a node of variable $?x$ with a heat value equal to 1. By this process, a heat query would be bigger in size with more workload queries getting combined regardless of their order in the workload.

6   Elements of Adaption

In this section, we describe the performance benefits of the indexes, join cache, and replication. We use these benefits besides the access rates and the storage cost to perform our universal adaption in Sect. 7.

6.1   Indexes

In any typical key-value RDF store, the data resides basically in indexes. An index is implemented as a hash table. For instance, the SPO index stores all the triples by hashing them on the subject. We can get a set of all triples that match a given subject by one lookup operation. This set is ordered on the predicate then on the object. Thus we can search that set for a certain predicate in logarithmic time. That index is optimal to answer any triple pattern[1] in the form $(s_1, p_1, ?o_1)$. However, if the triple pattern is in the form $(s_1, ?p_1, o_1)$, then we can still use SPO to answer it, but with an extra linear search for o_1 within the set returned from the hash-table lookup operation on s_1. We refer to this extra cost as *scanLookupTime* and is recorded in the heat query as performance benefits to the missing index. Another performance benefit of indexes comes from join. A SPARQL query is typically composed of multiple triple patterns that require join operation. The join planner selects a potential optimal join plan given that it implements all the possible 6 indexes. However, in case some indexes are missing, the query can still be executed using a sub-optimal join plan (i.e. the best plan that uses only available indexes). We refer to the cost difference between the two plans as *treeTime*. That cost is considered as performance benefits accounted for the missing indexes for the triple patterns been joined.

The performance benefit of a triple pattern t in an index χ is then the summation of the both given benefits:

$$\eta_{idx}(t, \chi) = scanLookupTime(t, \chi) + treeTime(t, \chi) \qquad (2)$$

6.2   Join Cache

Besides the six basic indexes, a triple store can have more cache-indexes to speed up the join process. However, those indexes require even more storage space. Fortunately, our adaptive system can measure and compare the cost-effectiveness of the cache indexes with other indexes and choose to build them only for the data that delivers higher benefits with respect to other indexes. We use two cache indexes: PPX and *typeIndex*. The PPX is hashed on two predicates. Given that the predicate is mostly constant in any triple pattern [9], PPX can store the results of almost any two joined patterns. The *typeIndex* is especially helpful in the queries that heavily use the predicate "*type*" like the LUBM benchmark [18]. Consider for instance the query: *(?x :graduatedFrom ?y. ?x :type :student. ?y :type :university)*. Its result can be cached in *typeIndex* with the key *(:student,:graduatedFrom,:university)*, and the benefit is the value of the saved join cost. Similar to the basic indexes, our system records the benefit and the usage values of the cache indexes in the heat queries and retrieves them at the adaption time.

6.3   Replication

In a federated distributed triple store with n working nodes, the system needs to generate at least n partitions out of the global RDF graph. For this purpose, a graph-based partitioning approach is widely used based on graph min-cut algorithms. The aim is to decrease the communication cost among the resulted partitions [8, 11, 12, 19, 21]. However, the problem is shifted to the border region where the edges are connecting multiple partitions. That problem is overcome

---

[1] According to SPARQL syntax $?o_1$ is a variable and others are constants or literals.

by replication in [11,12] at the cost of more storage space. Our adaptive system integrates the replication decision with the indexes by modeling it in the cost model Sect. 4. For that, we need the benefit, access rate, and size. The performance benefit of replicating a block of triples is saving the communication cost of transferring the block over the network. Such a cost is related to the network speed and status. We derive the access rate of replications using the workload analysis module explained in Sect. 5. From the perspective of a certain working node i, the general access rate to a remote triple is related to its distance from the border. This can be formulated in the following:

$$p_{rem}(v,i) = \frac{1}{outdepth(v,i)} \cdot p_{border} \tag{3}$$

Where $outdepth(v,i)$ is the distance to reach vertex v (minimum number of hops) from the border of partition i, and p_{border} is the probability of a query at partition i to access its border region. The value of p_{border} is initially set to 1, but is going to be further updated depending on the workload by counting the rate of accessing the border region by all the executed queries in the system so far. This value is related to the average length of the query. The longer the query the more its probability to touch the border region. Next, we derive a specific access rate for replication, again from the heat queries. This is achieved by recording the replication usage rate for each triple pattern which requires border data. That allows the heat query to extend to the remote data over the border region. Finally, the aggregation phase takes place. Any remote triple that resides in index χ gets the following aggregate access value from node i:

$$repAccess(v,i) = p_{rem}(v,i) + heatQueryAccess(v,\chi) \tag{4}$$

7   Universal Adaption

We have formulated the cost model of the indexes, replications, and join cache in terms of benefits, access rates, and storage costs. Our optimizer builds their statistics during query execution in *stat phase*. It performs an *adaption phase* using Algorithm 1 at each node. The input to the algorithm is two sets of operation rules. We define an operation rule in the following:

Definition 1 (Operational Rule). *An operational rule is defined as $\varpi_{op} = (\chi, s, a, \Delta)$, where χ is an index, s is a set of patterns that defines a set of triples D, a and Δ are functions that assign an access and benefit values to each $d \in D$. According to Formula 1, the benefit of each source can be calculated. We refer to $\bar{s}$ as the source with the maximum benefit in s, and:*

$$b(\varpi_{op}) = \frac{a(\bar{s}) \cdot \Delta(\bar{s})}{size(\bar{s})}$$

We define one rule for each basic and cache index for the local data, and another for the replicated data. Each rule is put in the assigned rules set R_a and the proposed rules set R_p. R_a represents what is already assigned to memory and R_p represents the proposed. The algorithm starts by a loop which first process the workload stats stored in the heat queries in Line 2, and updates the access functions of the rules accordingly. Line 3 sorts ascendingly the patterns of each $r_a \in R_a$ by relative benefits such that $\bar{s}$ is the pattern at the top. The sort is descendingly for R_p. The relative benefit is calculated using Formula 1 and Definition 1. Line 6 and 7 find the worst rule from R_p and the best from R_a, and swap them in Line 13. The algorithm terminates in Line 9, when the benefit of the best assigned rule is higher than of the worst of the proposed. The time required for the sort operation is bounded by the number of patterns which is limited. Most of the algorithm time is spent on evaluating the patterns and swapping the triples. However, the adaption phase takes place only when the system is idle and has no query to execute.

Algorithm 1: Rules-based space adaption algorithm

---

    input : RDF graph $G = \{V, E\}$, heat queries set H and two sets of the system operational
              rules: proposed rules R_p and assigned rules R_a

1   **for** *each* $r \in R_p \cup R_a$ **do**
2 | $r \leftarrow$ updateRulesAccess(r, H);
3 | $r \leftarrow$ sortRuleBySource(r);
4   **end**
5   **while** *true* **do**
6 | $r_p \leftarrow \varpi_{op} | \varpi_{op} = (\chi, s_p, a_p, \Delta_p) : \forall r_i \in R_p, b(\varpi_{op}) \geq b(r_i)$;
7 | $r_a \leftarrow \varpi_{op} | \varpi_{op} = (\chi, s_a, a_a, \Delta_a) : \forall r_i \in R_a, b(\varpi_{op}) < b(r_i)$;
8 | **if** $b(r_a) \geq b(r_p)$ **then**
9 | | **break**
10 | **end**
11 | $\hat{V}_p \leftarrow$ evaluate$(\bar{s}_p)$;
12 | $\hat{V}_a \leftarrow$ evaluate$(\bar{s}_a)$;
13 | swapAssignment$((r_p, \hat{V}_p), (r_a, \hat{V}_a))$;
14   **end**

---

8   Evaluation

In this section, we evaluate the universal adaption approach implemented in UniAdapt. We use the LUBM [18] is a generated RDF data set that contains data about universities. The size of the generated data is over 1 billion triples. The benchmark contains 14 test queries[2] that are labeled L_1 to L_{14}. The queries can be classified into 2 categories: bounded and unbounded. The bounded queries contain at least one constant vertex (subject or object) excluding the triple pattern that has "type" as predicate. The unbounded queries are L_2 and L_9, and all the others are bounded. The execution of a bounded query is limited to certain part of the RDF graph. Those parts can be efficiently recognized during query execution given the existence of a proper index.

---

[2] http://swat.cse.lehigh.edu/projects/lubm/queries-sparql.txt.

No Storage Limit. Since the systems used in comparison do not have space adaption, we run this part of the tests under no storage restrictions. We "train" the systems by an initial workload of 100 queries similar to the 14 queries of LUBM. Then we run a batch that contains 64 repetitions of each query on the systems shown in Table 1, and record the average runtime for the given queries. There was a clear superiority in the performance regarding the unbounded queries L_2 and L_9, and the bounded query L_8. However, the other bounded queries L_3 to L_6 behaved very closely for both UniAdapt and AdPart due to their relative simplicity. For the bounded queries L_4, L_6, L_7 and L_8, the running times were generally much less than the previous unbounded queries (Table 2). Both systems performed closely for L_4 and L_6. But Unidapt was superior in adapting to L_8, while AdPart showed better performance to L_7.

Table 1. Runtimes (ms) of LUBM queries

	L_2	L_3	L_4	L_5	L_6	L_7	L_8	L_9
UniAdapt	178	1	1	1	5	45	10	347
AdPart	7K	4	2	3	4	88	370	1K
H-RDF-3X	12K	3K	3K	3K	3K	7K	5K	9K
H-RDF-3X+	11K	3K	3K	3K	3K	5K	5K	8K

Table 2. Runtimes (ms) of bounded queries with 2 batches (b_1 and b_2)

	L4		L6		L7		L8	
	b_1	b_2	b_1	b_2	b_1	b_2	b_1	b_2
UniAdapt	2	1	5	4	45	15	10	1
AdPart	3	1	4	4	88	.1	370	1.5

Storage-Workload Adaption. In this part, we evaluate the unique adaption capabilities of UniAdapt with the storage space and the workload. In this context, we set three levels of storage capacity: $S_{2.5}$, S_5, and S_7. With capacity S_5 the system can potentially maintain 5 full indexes but may also decide to use the free space for join cache or replication. Moreover, we

Table 3. Generated workload properties of LUBM data set

Workload	Bounding	Length	Distribution
WLu1	No	3–4	Uniform
WLu2	No	3–4	50% to 50%
WLu3	No	3–4	90% to 10%
WLb1	Yes	3–4	Uniform
WLb2	Yes	3–4	50% to 50%
WLb3	Yes	3–4	90% to 10%

generated six types of workload from the LUBM data-set that are given by Table 3. The fact that all the queries have the "type" predicates allowed the UniAdapt to maintain only two indexes that are type indexes. One is sorted on the subject while the other is sorted on the object. The first run of the WLu1 in the space capacity of $S_{2.5}$ took relatively long (Fig. 2), due to the lack of the OPS or SPO indexes that are used to decrease the cost of further join; besides the difficulties to perform enough replications on the limited space to save the communication cost. The uniform workload distribution amplifies the problem by hardening the task of the optimizer to detect highly accessed data using its specific rules. Moving from $S_{2.5}$ to S_5 allows the system to have enough replications, as well as two more indexes besides its two type indexes. This is reflected in an obvious decrease in the execution time. Upgrading the capacity to S_7, provides enough space for more and better cache. That showed the highest decrease in the query execution time. In addition, moving towards better workload quality helped the specific rules to better detect the highly accessed data. This is seen when moving from WLu1, WLu2 to WLu3. The workload impact was much higher on the bounded queries in workloads WLb1, WLb2, and WLb3. Starting with WLb1 at capacity level of $S_{2.5}$ caused a high increase of the query execution time. This is because the space is only enough for two full indexes. Increasing the space to S_5 relaxed the optimization process and allowed the system to have both OPS and SPO indexes for most of the data, and overrides the issue of the low workload quality.

However, given a low workload quality the system achieves a useful cache only for storage level of s_7. Moving towards the better-quality workload WLb2 resulted in a high decrease in the execution time, even for the limited storage level of $s_{2.5}$, as the system is able to detect the hotter parts of the data

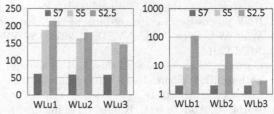

Fig. 2. Adaption with workload and space (average running time per query in milliseconds

using the specific rules. Moving to the excellent workload of WLb3 flattened the differences between the storage levels, as most of the workload is now targeting a very small region of the data, which can be efficiently indexed, replicated, and cached using the specific rules. That results in a high boost to the performance.

9   Conclusion and Future Work

In this paper, we presented the universal adaption approach for the storage layer of a distributed RDF triple store called UniAdapt. The adaption process aims to adapt the limited storage to store the most beneficial data within indexes, replication, and join caches. We defined a dynamic cost model that engages the benefit of each data assignment with its usage rate as well as its storage cost. Our experimental results showed the impact of the universal adaption on the query execution in different levels of storage space and workload quality. UniAdapt was able to excel in difficult levels of storage space capacity. On the other hand, it was able to flexibly adapt to space abundance for more query performance. For a future work, we consider adding the temporal effect of the workload in the heat query structures.

References

1. Aluc, G., Özsu, M.T., Daudjee, K.: Workload matters: why RDF databases need a new design. Proc. VLDB Endow. **7**(10), 837–840 (2014)

2. Aluç, G., Özsu, M.T., Daudjee, K.: Building self-clustering RDF databases using tunable-LSH. VLDB J. **28**(2), 173–195 (2019)
3. Ceri, S., Negri, M., Pelagatti, G.: Horizontal data partitioning in database design. In: SIGMOD Conference, pp. 128–136. ACM Press (1982)
4. Chong, E.I., Das, S., Eadon, G., Srinivasan, J.: An efficient SQL-based RDF querying scheme. In: VLDB, pp. 1216–1227. ACM (2005)
5. Dasgupta, S., Papadimitriou, C.H., Vazirani, U.V.: Algorithms. McGraw-Hill, New York (2008)
6. Erxleben, F., Günther, M., Krötzsch, M., Mendez, J., Vrandečić, D.: Introducing wikidata to the linked data web. In: Mika, P., et al. (eds.) ISWC 2014. LNCS, vol. 8796, pp. 50–65. Springer, Cham (2014). https://doi.org/10.1007/978-3-319-11964-9_4
7. Galárraga, L., Hose, K., Schenkel, R.: Partout: a distributed engine for efficient RDF processing. In: Proceedings of the 23rd International Conference on World Wide Web (New York, NY, USA, 2014), WWW 2014 Companion, pp. 267–268. ACM (2014)
8. Gurajada, S., Seufert, S., Miliaraki, I., Theobald, M.: TriAD: a distributed shared-nothing RDF engine based on asynchronous message passing. In: SIGMOD, pp. 289–300. ACM (2014)
9. Harbi, R., Abdelaziz, I., Kalnis, P., Mamoulis, N., Ebrahim, Y., Sahli, M.: Accelerating SPARQL queries by exploiting hash-based locality and adaptive partitioning. VLDB J. **25**(3), 355–380 (2016)
10. Hayes, P.: RDF semantics, W3C Recommendation 10 February (2004). https://www.w3.org/TR/rdf-mt/
11. Hose, K., Schenkel, R.: WARP: workload-aware replication and partitioning for RDF. In: ICDE Workshops, pp. 1–6. IEEE Computer Society (2013)
12. Huang, J., Abadi, D.J., Ren, K.: Scalable SPARQL querying of large RDF graphs. Proc. VLDB Endow. **4**(11), 1123–1134 (2011)
13. Karypis Lab: METIS: family of graph and hypergraph partitioning software (2020). http://glaros.dtc.umn.edu/gkhome/views/metis
14. Moerkotte, G., Neumann, T.: Analysis of two existing and one new dynamic programming algorithm for the generation of optimal bushy join trees without cross products. In: VLDB, pp. 930–941. VLDB Endowment (2006)
15. Monaci, M., Pferschy, U., Serafini, P.: Exact solution of the robust knapsack problem. Comput. Oper. Res. **40**(11), 2625–2631 (2013)
16. Neumann, T., Weikum, G.: The RDF-3X engine for scalable management of RDF data. VLDB J. **19**(1), 91–113 (2010)
17. Peng, P., Zou, L., Chen, L., Zhao, D.: Query workload-based RDF graph fragmentation and allocation. In: EDBT, pp. 377–388. OpenProceedings.org (2016)
18. SWAT Projects: The Lehigh University Benchmark (LUBM). http://swat.cse.lehigh.edu/projects/lubm/
19. Wang, L., Xiao, Y., Shao, B., Wang, H.: How to partition a billion-node graph. In: ICDE, pp. 568–579. IEEE Computer Society (2014)
20. Weiss, C., Karras, P., Bernstein, A.: Hexastore: sextuple indexing for semantic web data management. Proc. VLDB Endow. **1**(1), 1008–1019 (2008)
21. Zhang, X., Chen, L., Tong, Y., Wang, M.: EAGRE: towards scalable I/O efficient SPARQL query evaluation on the cloud. In: ICDE, pp. 565–576. IEEE Computer Society (2013)

RDF Data Management is an Analytical Market, not a Transaction One

Olivier Curé[1]([✉]), Christophe Callé[1,2], and Philippe Calvez[2]

[1] LIGM Univ. Paris Est Marne la Vallée, CNRS, 77454 Marne la Vallée, France
{olivier.cure,christophe.calle}@univ-eiffel.fr
[2] ENGIE LAB CRIGEN, Saint-Denis, France
philippe.calvez1@engie.com

Abstract. In recent years, the Resource Description Framework data model has seen an increasing adoption in Web applications and IT in general. This has contributed to the establishment of standards such as the SPARQL query language and the emergence of production-ready database management systems based on this data model. In this paper, we however argue that by concentrating on transaction related functionalities rather than analytical operations, most of these systems address the wrong data market. We motivate this claim by presenting several concrete arguments.

1 Introduction

The Resource Description Framework (RDF) data model has attracted lot of attention during these last few years. It enabled the design and implementation of many Web and IT applications. For instance, it supports innovative approaches for artificial intelligence's knowledge representation and web search. This is due to the publication of some of the largest and most popular knowledge graphs (KG), *e.g.*, DBpedia, Wikidata, Bio2RDF, UniProt, which are represented using this data model. As such, RDF is now recognized has one of the leading data model in the graph database management ecosystem. Compared to its direct competitor, the Labelled Property Graph (LPG) data model, RDF presents great support for data integration and reasoning services. These two features are due (i) to the omnipresence of Internationalized Resource Identifiers (IRI) in RDF graphs which serve as a common identifying solution at the scale of the Web and (ii) to associations with semantically-rich vocabularies, generally denoted ontologies, which together with reasoners enable to compute inferences.

This increase of interest for RDF has led to the emergence of efficient, production-ready RDF stores (see Table 1's first column for a list of the most prominent systems). In addition to natively providing reasoning services in different ontology languages, these systems possess some important functionalities that one can expect from a standard relational database management system (DBMS), *e.g.*, optimized query processing for a declarative language (typically SPARQL). These functionalities also include on line transaction process-

M. Golfarelli et al. (Eds.): DaWaK 2021, LNCS 12925, pp. 109–115, 2021.
https://doi.org/10.1007/978-3-030-86534-4_9

ing (OLTP) through the support of ACID properties. Hence, they have been designed to process high rates of (update) SPARQL queries per second.

In this paper, we argue that the support for ACID transactions is not frequently used in RDF stores. This is mainly due to the fact it is not being considered as a critical feature for most RDF-based application development. We motivate this claim in Sect. 2.1 from technical and contextual aspects as well as interviews conducted with project leaders. In Sect. 2.2, we highlight that analytical operations, not generally present in RDF stores, are highly expected in many applications. In Sect. 3, we present two general forms of analytics and emphasize that one is more needed than the other. Finally, we conclude the paper and present some perspectives.

Table 1. RDF stores characteristics, considering reasoning, RDFS:r, RDFS+: r+, OWL Lite: l n OWL QL: q, OWLRL: d, OWL Horst: h, OWLDL: dl)

Name	Transaction (ACID)	Reasoning	Analytic features	Materialized views
AllegroGraph[a]	Yes	r+, d	No	No
AnzoGraph DB[b]	Yes	r+	Graph analytics + OLAP style	Yes
Blazegraph[c]	Yes	r+, l	No	No
GraphDB[d]	Yes	r, q, d, h	No	No
MarkLogic[e]	Yes	r, r+, h	No	No
Oracle[f]	Yes	r, q, d	No	No
RDFox[g]	Yes	r+, d	No	No
Stardog[h]	Yes	All	No	No
Virtuoso[i]	Yes	r+	No	No

[a] https://franz.com/agraph/allegrograph/.
[b] https://www.cambridgesemantics.com/anzograph/.
[c] https://blazegraph.com/.
[d] https://www.ontotext.com/products/graphdb/.
[e] https://www.marklogic.com/.
[f] https://www.oracle.com/.
[g] https://www.oxfordsemantic.tech/product.
[h] https://www.stardog.com/.
[i] https://virtuoso.openlinksw.com/.

2   Arguments for OLAP RDF Stores

2.1   Why RDF Stores Do Need ACID Transactions?

In this section, we motivate the fact that many RDF-based applications do not require full ACID transaction guarantees.

SPARQL and Update Operations. Since 2008 and its first W3C recommendation release, SPARQL is the established query language for RDF data. In 2013, a set of new features to the query language were added, leading to the SPARQL 1.1 recommendation. Among these features, update operations were introduced. This means that for over five years, end-users had to either programmatically update RDF data sets or use a non-standard, *i.e.*, system specific, declarative update solution. In a data management market anchored in transaction processing is seems unreal to leave application developers in such a situation for over five years. Hence, one can doubt that OLTP is really the market for RDF stores. One obvious question is: are the workload generally dealt with in RDF stores transactional in nature?

Linked Data Update Frequency. The popularity of RDF is partly due to its ability to integrate data and thus to break data silos. This is mainly made possible by linking these silos based on relating nodes of different graphs. The Linked Data movement and the creation of very large KGs are the most concrete examples of such an approach. The Linked Open Data cloud[1] currently connects over 1.200 graphs with an estimation of tens of billions triples over domains as diverse as life sciences, media, social networking, government, etc. Some of these KGs are produced out of extractions from open data repositories and Web scraping from Web sites. Although these sources, *e.g.*, Wikipedia, are updated at a per second rate, popular KGs are generally updated in term of days to weeks, *e.g.*, a release frequency for DBpedia, UniProt and Wikidata occurs respectively every 6 months, 4 weeks and every couple of days. So, we are far from the kind of transaction rates that ACID-compliant DBMS can support, *e.g.*, in the range of several thousands to millions of transactions per second. In fact, we are closer to the bulk loading approach of data warehouses where the main purpose of the system is to analyze data sets rather than manage the freshest data. We can then ask ourselves what is inherently complex about executing updates on an RDF graph?

Reasoning Issues. Together with data integration, the main benefit of using the RDF data model for a graph database is to benefit from reasoning services. These services depend on ontologies that can be defined with more or less expressive ontology languages (from RDFS to OWLDL to at least stay within decidable fragments). As displayed in the 'reasoning' column of Table 1, most production-ready RDF stores address rather low ontology expressiveness. This is mainly due to the practical cost of computing inferences which is already high for the least expressive languages, *e.g.*, RDFS entailment is a NP-complete problem [6]. In RDF stores, reasoning is commonly addressed via either a materialization or a query rewriting approach. In the former, whenever some updates are submitted to the DBMS, a reasoner computes the corresponding inferences and updates the stored graph accordingly. We can easily understand that this approach, although efficient considering query processing, has its limitations when the rate of updates is high. In fact, for any updates, the system needs to check whether some inferences can be deduced. So reasoning is invoked for each update operation. Incremental reasoning as proposed in RDFox [4] potentially improves the cost of materialization but it nevertheless does not enable high rate

---

[1] https://lod-cloud.net/.

of update transactions. In the latter approach, *i.e.*, query rewriting, reasoning is performed at query run time. Hence, handling updates is much more efficient than in the materialization approach but query processing is much less efficient. For this reason, more systems are adopting materialization than query rewriting, but some systems adopt both (*e.g.*, Allegrograph). We can thus understand that computing inferences at the rate of tens to hundreds of transactions per second is not realistic.

Real-World Use Cases. During the last few years, we have conducted several interviews and discussions with large organizations that are intensively managing RDF databases of up to several TB, *e.g.*, Publication Office of the European Union, French ministry of culture, institutes dealing with sensus in France and Italy. We found out that these companies are not updating their databases through high transaction rates but are rather bulk loading data in manner reminiscent to data warehouses, *i.e.*, at a per day or per week frequency rate. The main reasons for this update pattern is mainly due to the cost of live reasoning over incoming triples. Moreover, the companies that were not using any form of reasoning on RDF data, while their ontologies would permit to, were not using high transaction rates.

2.2   RDF Stores Should Support Analytical Operations

We now provide some arguments toward enriching RDF stores with analytical operations. These arguments come from the companies and end-users of RDF stores.

Use Cases Emphasized by RDF Sellers. We have collected the list of the most commonly encountered use cases of Table 1's RDF stores. Since, these systems have all commercial editions (Blazegraph is now AWS Neptune), we can believe that the use cases correspond to the needs of theirs customers. Among the top nine entries, we are finding the following: Advanced search and discovery, analytics/Business Intelligence, fraud detection and recommendations. These tasks are clearly relevant in analytical oriented DBMS. It is interesting to note that none of the other common use cases are really requiring ACID transaction guarantees but are rather related to data integration or smart metadata management.

End-User Point of Views. VLDB 2017's best paper [5] provides a nice survey on the usage of graph processing and graph data management system (for both LPG and RDF data models). It highlights that analytics is the task where end-users are spending the most hours (more than testing, debugging, maintenance, ETL and cleaning). Moreover, the top graph computations performed on these systems, DBMS including, are finding connected components, neighborhood queries, finding shortest paths, subgraph matching (*i.e.*, SPARQL), ranking and centrality scores, reachability queries. These operations are quite useful in queries performing some kind of recommendations, *e.g.*, in medicine to identify a certain molecule or in culture to find a popular artist or art form.

3   The Road to Analytics in RDF Stores

3.1   Kind of Analytical Operations

Two forms of analytical operations can be considered for RDF stores. In the former, the idea is to adapt the multidimensional aspects that we can find in relational OLAP systems (ROLAP). They enable the management of information cubes and are generally extending the SQL query language with new operations such as roll-up, drill-down, pivot, slice and dice. In order to facilitate this management, they rely on a specific kinds of schemata that organize database tables in a certain manner. These schemata correspond to so-called star, snowflake or constellation. The adaptation of this approach to the RDF data model is not straightforward due to the schemaless nature of RDF, and graph models in general (*e.g.*, LPG). Moreover, in these schemata, the fact table(s) is (are) supposed to store data originating from OLTP systems. Although these data can be bulk loaded in the RDF stores, we consider that it would be better to leave them in the relational DBMS and organize some linkage solution with the RDF stores in the style of virtual KG [7]. The low frequency of update operations on ROLAP systems favors the use of materialized views (as opposed to virtual views of OLTP). Such views can be useful in RDF stores where it can impact query processing. They can be created via the SPARQL CONSTRUCT query form and hence enable RDF compositional queries. The work presented in [1] is an attempt to integrate these cube operations in SPARQL and RDF data management. The same researchers have implemented the SPADE [2] system on top of this approach but the code is not available and the approach does not seem to have been adopted in existing systems.

The second form of analytical operations correspond to graph algorithms. Until recently, they were generally present in large graph processing systems, *i.e.*, GraphX, Giraph, Gelly, but definitely make sense in a graph oriented DBMS where the data management is given much more attention. These graphs algorithms correspond to finding connected components and shortest paths, getting centrality and ranking scores, counting and enumerating connected components. Again, these algorithms are quite relevant for RDF stores since they meet the high expectations of typical end-users.

3.2   Emerging Systems

One DBMS adopting the RDF data model has in fact started to consider the analytical direction proposed in this paper. It corresponds to AnzoGraph DB which happens to be the most recent store among our list of production-ready system (only RDFox is more recent). AnzoGraph DB proposes a complete support of

SPARQL 1.1 with RDFS+ and OWLRL inferencing (via triple materialization), OLAP style analytics with windowed aggregates, cube, rollup, grouping sets and large set of functions. Moreover, it supports graph algorithms such as page rank, betweenness centrality, connected components, triangle enumeration, shortest and all paths. To the best of our knowledge, AnzoGraph DB is the only DBMS to support materialized views.

Among other production-ready RDF stores, we see the first movement toward analytics. For instance, Stardog in its 7.5 release (january 2021) provides a beta graph analytics component which contains operations such as page rank, label propagation, (strongly) connected components. Finally, outside of pure RDF store players, a similar trend is catching up with SANSA [3]. It is defined by its creators as a "big data engine for scalable processing of large-scale RDF data" and takes the form of a set of libraries. It is able to perform reasoning, querying and analytic operations. Considering analytics, it relies on either Apache Spark[2] or Apache Flink[3] distributed computing frameworks. Hence, it benefits from their respective GraphX and Gelly graph components to compute graph algorithms. Nevertheless, this can only be specified by mixing some SPARQL queries with some programs compiling over a Java environment.

4   Conclusion

In this vision paper, we have motivated the fact that the playground of RDF stores consists more of analytical than transactional processing. The limitation toward processing high rates of transaction mainly lies in the cost of inferences, the schemaless characteristic of RDF data considering multidimensional-based analytics. We believe that analytical operations based on standard graph processing are the most relevant for end-users. Potentially, some very interesting features should emerge for RDF analytical DBMS, *e.g.*, mixing reasoning with analytical operations and analytic-based data integration. Obtaining such services will come at the price of providing a seamless collaboration between relational DBMS for managing transactional data and RDF stores for computing graph analytics.

Finally, a similar trend is occurring in systems based on the LPG data model. Among the plethora of production-ready systems, *e.g.*, Neo4J, JanusGraph, RedisGraph, currently only TigerGraph seems to consider graph analytics as a promising market.

---

[2] https://spark.apache.org/.
[3] https://flink.apache.org/.

References

1. Colazzo, D., Goasdoué, F., Manolescu, I., Roatis, A.: RDF analytics: lenses over semantic graphs. In: Chung, C., Broder, A.Z., Shim, K., Suel, T. (eds.) 23rd International World Wide Web Conference, WWW 2014, Seoul, Republic of Korea, 7–11 April, 2014, pp. 467–478. ACM (2014)
2. Diao, Y., Guzewicz, P., Manolescu, I., Mazuran, M.: Spade: a modular framework for analytical exploration of RDF graphs. Proc. VLDB Endow. **12**(12), 1926–1929 (2019)
3. Lehmann, J., et al.: Distributed semantic analytics using the SANSA stack. In: d'Amato, C., et al. (eds.) ISWC 2017. LNCS, vol. 10588, pp. 147–155. Springer, Cham (2017). https://doi.org/10.1007/978-3-319-68204-4_15
4. Nenov, Y., Piro, R., Motik, B., Horrocks, I., Wu, Z., Banerjee, J.: RDFox: a highly-scalable RDF store. In: Arenas, M., et al. (eds.) ISWC 2015. LNCS, vol. 9367, pp. 3–20. Springer, Cham (2015). https://doi.org/10.1007/978-3-319-25010-6_1
5. Sahu, S., Mhedhbi, A., Salihoglu, S., Lin, J., Özsu, M.T.: The ubiquity of large graphs and surprising challenges of graph processing. Proc. VLDB Endow. **11**(4), 420–431 (2017)
6. ter Horst, H.J.: Completeness, decidability and complexity of entailment for RDF schema and a semantic extension involving the owl vocabulary. Web Semant. **3**(2–3), 79–115 (2005)
7. Xiao, G., Ding, L., Cogrel, B., Calvanese, D.: Virtual knowledge graphs: an overview of systems and use cases. Data Intell. **1**(3), 201–223 (2019)

Document Ranking for Curated Document Databases Using BERT and Knowledge Graph Embeddings: Introducing GRAB-Rank

Iqra Muhammad[1]([✉]), Danushka Bollegala[1], Frans Coenen[1], Carrol Gamble[2], Anna Kearney[2], and Paula Williamson[2]

[1] Department of Computer Science, The University of Liverpool, Liverpool L69 3BX, UK
iqra.muhammad@liverpool.ac.uk
[2] Department of Biostatistics, Institute of Translational Medicine, The University of Liverpool, Liverpool L69 3BX, UK

Abstract. Curated Document Databases (CDD) play an important role in helping researchers find relevant articles in scientific literature. Considerable recent attention has been given to the use of various document ranking algorithms to support the maintenance of CDDs. The typical approach is to represent the update document collection using a form of word embedding and to input this into a ranking model; the resulting document rankings can then be used to decide which documents should be added to the CDD and which should be rejected. The hypothesis considered in this paper is that a better ranking model can be produced if a hybrid embedding is used. To this end the Knowledge Graph And BERT Ranking (GRAB-Rank) approach is presented. The Online Resource for Recruitment research in Clinical trials (ORRCA) CDD was used as a focus for the work and as a means of evaluating the proposed technique. The GRAB-Rank approach is fully described and evaluated in the context of learning to rank for the purpose of maintaining CDDs. The evaluation indicates that the hypothesis is correct, hybrid embedding outperforms individual embeddings used in isolation. The evaluation also indicates that GRAB-Rank outperforms a traditional approach based on BM25 and a ngram-based SVR document ranking approach.

Keywords: BERT · Knowledge graph concepts · Document ranking

1 Introduction

The number of published papers in scientific research is increasing rapidly in any given domain. Consequently, researchers find it difficult to keep up with the exponential growth of the scientific literature. In order to address this challenge many organisations manage Curated Document Databases (CDDs). CDDs are

© Springer Nature Switzerland AG 2021
M. Golfarelli et al. (Eds.): DaWaK 2021, LNCS 12925, pp. 116–127, 2021.
https://doi.org/10.1007/978-3-030-86534-4_10

specialised document collections that bring together published work, in a defined domain, into a single scientific literature repository. One example of such a CDD, and that used for illustrative purposes in this paper, is the Online Resource for Recruitment research in Clinical trials (ORRCA[1]) CDD [7]. The ORRCA CDD brings together abstracts of papers concerned with the highly specialised domain of recruitment strategies for clinical trials.

The provision of CDDs provide a useful facility for researchers. However, for CDDs to remain useful, they must be constantly updated, otherwise their utility is of only temporary value. The challenge is illustrated in the context of the ORRCA CDD in Fig. 1. From the figure the exponential, year-on-year, growth of the number of papers can be observed clearly. Updates are conducted using what is referred to as a *systematic review process*. The systematic review is typically conducted manually by querying larger document collections, a time consuming task. In the case of ORRCA, the PubMed search engine for the MEDLINE life sciences and biomedical abstracts database was used for the systematic review. The process can be enhanced using *document ranking* so that candidates for an update can be ranked according to relevance and the top k considered in more detail, whilst the remainder can be rejected.

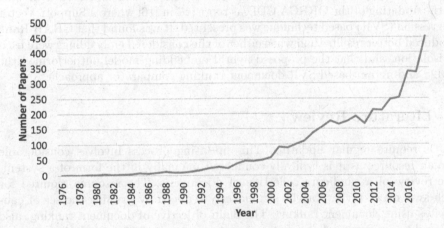

Fig. 1. ORRCA papers and articles 1976–2017, illustrating the exponential growth of the number of publications directed at recruitment strategies for clinical trials.

Document ranking has been extensively used in the context of document retrieval. The traditional approach, given a particular search query, is to rank documents using a frequency measure that counts the frequency whereby terms in the search query appear in each candidate document [6,18]. However, frequency based document ranking models fail to capture the semantic content behind individual search queries. An alternative is to use a learning to rank model more suited to capturing the semantic meaning underpinning search

---
[1] https://www.orrca.org.uk/.

queries [14,16]. Recent work on Learning to Rank (LETOR) has used word embeddings of various kind as the input [11,14]. Word embeddings can be learnt from scratch or a pre-trained embedding model can be adopted. Existing word embedding based approaches to LETOR have focused on a single embedding, with good results; a popular choice is to use Bidirectional Encoder Representations from Transformers (BERT) embeddings [10,17]. The intuition presented in this paper is that a *hybrid approach* using two orthogonal, but compatible, embeddings will result in a more effective ranking (in the context of the CDD update problem). An intuition that is supported by the observations given in [1]. To this end this paper presents the Knowledge Graph And BERT Ranking (GRAB-Rank) approach to LETOR, designed to support the periodic updating of CDDs, that combines BERT word embeddings and knowledge graph concept embeddings; the latter generated using a bespoke *random walk* technique.

The GRAB-Rank approach is fully described and evaluated. The proposed approach assumes the availability of a literature knowledge graph. Techniques whereby document knowledge graphs can be constructed, given a document corpus, are available (see for example [13]). The presented evaluation was conducted using the ORRCA CDD. GRAB-Rank results were compared with an approach based on the popular Okapi BM25 ranking function [23] and earlier work directed at the updating of the ORRCA CDD as reported in [16] where a Support Vector Regression (SVR) based technique was presented. It was found that GRAB-Rank produced better results than when either of the considered embeddings were used in isolation, and that the proposed hybrid embedding model outperformed the BM25 and ngram-based SVR document ranking comparator approaches.

2   Literature Review

CDDs require regular updating. This updating process involves considerable human resource as it is typically conducted manually in the form of a systematic review of a candidate collection of documents. The resource required for such systematic review can be significantly reduced by pruning the set of candidates using document ranking. The main objective of document ranking, also referred to as *score-and-sort*, is to compute a relevance score for each document and then generate an ordered list of documents so that the top k most relevant documents can be selected. In this paper a mechanism for updating the ORRCA CDD [7] is presented founded on a hybrid document ranking technique. Recent work in document ranking has been focused on using external knowledge for improving document rankings. Especially the use of contextualised models such as BERT. LETOR models can be categorised as being either: (i) Traditional document ranking models, (ii) Semantic document ranking models, or (iii) Knowledge graph document ranking models. Subsections 2.1, 2.2 and 2.3 give further detail with respect to each of these categories.

2.1   Traditional Document Ranking Models

Traditional document ranking models are founded on statistical or probabilistic approaches. Many variants of these methodologies have been proposed and continue to be proposed. Most are founded on a vector space model of the input document corpus where the dimensions of the vector space are defined using terms that appear in the document collection. The terms to be included are typically selected using a scoring mechanism. Term Frequency - Inverse Document Frequency (TF-IDF) is a popular choice [11]. In this manner a n-dimensional vector space can be constructed. A popular algorithm for generating vector representations of words is GloVE (Global Vectors for Word Representation), an unsupervised learning algorithm that operates by aggregating global word-word co-occurrence statistics found in an input corpus [20]. An alternative mechanism of generating a vector space model is to use *word n-grams*. This was the technique used in [12] and [16]. The significance of the techniques used in [12] and [16] is that it was evaluated using the ORRCA CDD, and hence used with respect to the evaluation presented later in this paper to compare with the operation of GRAB-Rank. Once the input document corpus has been converted to a vector based representation, a document ranking function can be applied to rank the documents in decreasing order of relevance to a query. A popular ranking function is the Okapi BM25 ranking function which is founded on a probabilistic retrieval framework [23]. The Okapi BM25 function was also adopted as a document ranking baseline with respect to the work presented in this paper.

2.2   Semantic Document Ranking Models

The traditional statistical and probabilistic document ranking models assume each term is independent of its neighbours. Semantic document ranking models take into account the context of terms in relation to their neighbouring terms, in other words the "semantic" context associated with each term. We refer to this using the phrase *word embedding*. The distinction can be illustrated by considering the word "bank"; using a semantic context representation this would comprise a number of vectors depending on the context of the word "bank", either as: (i) an organisation for investing and borrowing money, (ii) the side of a river or lake, (iii) a long heap of some substance, (iv) the process of heaping up some substance or (v) the process of causing a vehicle to tilt to negotiate a corner. Using a non-contextualised representation the word "bank" would be represented using a single vector regardless of context.

Semantic representations are generated using a contextual model to generate the desired word embeddings; different terms that have the same semantic meaning are thus represented in a similar way. The required contextual model can be learned directly, typically using deep learning, from the document corpus of interest. Examples of document ranking systems that use a learnt contextual model to produce a word embedding can be found in [3,15,28]. However, learning a contextual model requires considerable resource. The alternative is to use an existing pre-trained contextual model to generate a word embedding for a given corpus. A popular choice of pre-trained contextual model is the Bidirectional Encoder Representations from Transformer (BERT). BERT takes into account the context of a target word using the surrounding words in a large corpora; BERT has been used with respect to many downstream natural language processing tasks including document ranking [15, 26]. An alternative contextual model that can be used is the embeddings from Language Model ELMo [21]. This model is based on deeply contextualized word embeddings which are created from Language Models (LMs). BERT is a transformer-based architecture while ELMo is Bi-LSTM Language model. BERT is purely Bi-directional and ELMo is semi-bidirectional. However, with respect to the work presented in this paper, because of BERT's popularity and its ease of use in Python, a BERT pre-trained model based sentence embeddings were used for the downstream task of ranking scientific abstracts.

2.3 Knowledge Graph Document Ranking Models

The work presented in this paper assumes CDDs represented as literature knowledge graphs. A knowledge graph is a collection of vertices and edges where the vertices represent entities or concepts, and the edges represent a relationship between entities and/or concepts. The reason for using knowledge graphs is that they provide efficient storage and retrieval in the context of linked descriptions of data. Some well known examples of knowledge graphs include Freebase [2] and YAGO [22]. In the context of document knowledge graphs the concepts stored at vertices represent semantic information which, it is argued here, can be used in the form of knowledge graph embeddings for document ranking purposes. Examples of recent work directed at knowledge graphs for document ranking include the entity-based language models described in [8,9,27]. This existing work has demonstrated the viability of knowledge graph based document ranking. The work presented in this paper proposes a hybrid approach that combines semantic document ranking with knowledge graph document ranking.

3 Problem Definition

A CDD is a data set of the form $D = \{d_1, d_2, \ldots d_n\}$ where each $d_i \in D$ is a document (research article/paper). For the CDD to remain useful it is necessary for it to be periodically updated by adding the set of recently published new documents Q to D so that $D_{new} = D \cup Q$. The set Q is traditionally generated using a systematic review process [7] applied to a larger data set U ($Q \subset U$). Systematic reviews involve a detailed plan and search strategy with the objective of identifying, appraising, and synthesizing all relevant studies on a particular topic [7,25]. The challenge is to automatically generate Q in such a way that Q contains as many relevant documents as possible. The anticipation is that it will not be possible to automatically generate a set Q that contains all relevant documents and no irrelevant documents. The idea is therefore to apply a Learning to rank model (LETOR) whereby U is ordered according to relevance score and the top k documents selected for potential inclusion in D.

4 BERT and Knowledge Graph Embeddings Based Document Ranking

This section presents the proposed Knowledge Graph and BERT Ranking (GRAB-Rank) approach. A schematic of the approach is presented in Fig. 2. The input is a collection of documents U to be potentially included in D. The next stage is to generate two sets of document embeddings: (i) document embeddings generated from a random walk of a knowledge graph G generated from U, and (ii) document embeddings generated using BERT. The first requires the transformation of U into G, how this can be achieved is presented in Subsect. 4.1. The process of generating document embeddings from G is then described in Subsect. 4.2. The process for generating document embeddings using BERT is described in Subsect. 4.3. Once we have the two types of document embeddings, these are combined into a single embedding, by concatenating one to the other. The concatenated embedding is then used as input to a LETOR model. With respect to the evaluation presented later in this paper, and as indicated in the figure, a Support Vector Regression (SVR) model was used to generate the document ranking. A SVR model was used because this has been shown to produce good results as evidenced in [12] and [16], a previously proposed approaches for updating CDDs which also focused on the ORRCA CDD. SVR uses the same principle as Support Vector Machines (SVMs) but with respect to regression problems. The SVR LETOR model, once learnt, can be used to assign a ranking value to each document in U. To obtain Q from U we then need a cut-off threshold value σ. The work in [12] reported the results from a sequence of experiments to establish the most appropriate value for σ. They found that 97% of relevant abstracts can be identified by considering the top 40–45% of potential abstracts. This was found to equate to a value of $\sigma = 0.30$. For the evaluation presented in this paper, $\sigma = 0.25$ was used (so as to include a "safety margin").

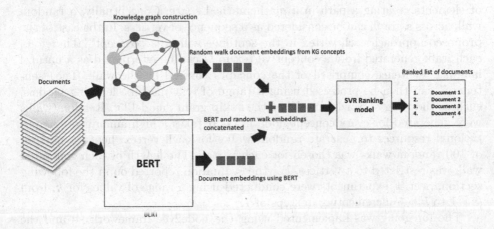

Fig. 2. Schematic of the GRAB-Rank approach.

4.1   Knowledge Graph Construction

The prerequisite of the GRAB-Rank approach for the maintenance of CDDs is a literature knowledge graph $G = \{V, E\}$ where the set of vertices V represent documents and concepts, and the set of edges E represent relationships between the vertices. There are various mechanisms whereby G can be constructed; it can be done manually, but is clearly better addressed in an automated manner. One proposed solution, and that adopted with respect to this paper, is the OIE4KGC (Open Information Extraction for Knowledge Graph Construction) approach presented in [13]. The OIE4KGC solution commences by extracting concept-relation-concept triples from a given document collection D (a CDD) using the RnnOIE Open Information Extraction (OIE) tool [24]. The triples are then filtered so that only the most relevant concepts are retained, each identified by a unique label. The retained triples, are then used to construct G such that the set of vertices V represents concepts and documents (in the following the terms concept vertex and document vertex are used to distinguish between the two), and the set of edges E comprises either: (i) the extracted relations from one concept to another concept, or (ii) "mention" relationships from a document to a concept. A mention relationship between a document and a concept implies that a document "mentions" this particular concept. Similar concepts in the knowledge graph were linked, using a biomedical entity linker.

4.2   Knowledge Graph Concept Embeddings Using a Random Walk

This section presents the process for generating concept embeddings from a literature knowledge graph (generated as described above); the first form of embedding in the proposed hybrid embedding approach. The process commences by generating a sequence of random walks (paths) linking concept vertices. The *random walk* idea was first proposed in [19], where it was defined as a sequence of elements creating a path in a mathematical space. Conceptually, a random walk across a graph can be considered as a sequence of vertices. In the case of the proposed approach each vertex in the sentence will be a "concept". Therefore, each walk generated from a concept vertex in G can be interpreted as a natural language sentence comprised of the concepts covered by the walk. The "sentences" can then be processed using a range of text machine learning models, such as the "bag of words" model or the "skip gram" model [5]. Random walks were generated for every concept node in G. It takes a high amount of computational resources to generate random walks for each vertex, hence a number of 100 random walks was chosen for each vertex. The length of each generated walk was restricted to k vertices. For the evaluation reported on in the following section, Sect. 4, experiments were conducted using a range of values for k, from $k = 1$ to $k = 5$ incrementing in steps of 1.

The foregoing was implemented using the node2vec framework [4] and the skip-gram model. Using the node2vec framework random walks can be generated using a number of strategies, these can be broadly categorised as Breadth-First Sampling (BFS) or Depth-First Sampling (DFS). The breadth-first strategy involves identifying all the immediate neighbours of a current vertex $v_i \in V$, to be included in the random walks to be generated and then moving on to immediate neighbours plus one, and so on until we reach random walks of length k. The depth-first strategy involves generating each entire random walk in turn rather than "in parallel". A breadth-first strategy (BFS) was used for the proposed GRAB-Rank approach.

4.3 BERT Contextualised Embeddings

This section discusses the contextualised embedding process, the second embedding used with respect to the proposed GRAB-Rank approach. The idea is to use transfer learning; the process of using a pre-trained deep language model to generate document embeddings. There are a number of such language models available, examples include ELMo [21] and BERT [15]. With respect to the evaluation presented later in this paper the BERT language model was used. The advantage offered by these models, as noted in Sect. 2, is that they are context aware; unlike many alternative models, such as GloVe [20], where each word is represented using a single vector regardless of context. Using a pre-trained language model, document embeddings are generated by replacing each word in a given document with the corresponding (BERT) word embedding. All the word vectors in the document are then concatenated to obtain a single document embedding. Contextualised language models consist of multiple stacked layers of representation (and an input layer); the greater (deeper) the number of layers the greater the extent of the context incorporated into a word representation. To generate word embeddings all layers can be used or the top n (most significant) layers. With respect to the evaluation presented later in this paper results are reported using all twelve BERT layers.

5 Evaluation

This section presents the evaluation of the proposed GRAB-Rank approach. For the evaluation, the abstracts for the 2015 and 2017 systematic review updates of the ORRCA CDD were used because: (i) a ground truth was readily available (the abstracts eventually selected for inclusion in ORRCA were known); and (ii) ORRCA had been used in previous LETOR studies, namely those presented in [12] and [16], hence a comparison could be conducted. Two data sets were generated using the 2015 and 2017 abstracts:

ORRCA-400. A small data set which could be manually inspected and analysed in the context of the proposed GRAB-Rank approach, referred to as the ORRCA-400 data set because it comprised 400 abstracts, 200 abstracts included in ORRCA and 200 excluded. Thus an even distribution,

ORRCA-Update. A much larger data set to test the scalability of the proposed approach made up of the entire 2015 and 2017 ORRCA update collections, 11,099 abstracts for the 2015 update (1302 included and 9797 excluded) and 14,485 for the 2017 update (1027 included and 13458 excluded).

Both datasets were pre-processed by removing punctuation and stop words. For stop word removal ntlk[2] was used. For training and testing a 60:40 training-testing split was used with respect to both data sets. Approaches similar to GRAB-Rank [10,15] have used similar splits for training and testing document ranking models.

The objectives of the evaluation were:

1. To conduct an ablation study to compare the operation of the proposed GRAB-Rank approach with using only BERT embeddings and only knowledge graph embeddings, so as to demonstrate that the proposed hybrid approach outperformed the component approaches when used in isolation.
2. To compare the operation of the proposed GRAB-Rank approach with alternative traditional document ranking systems applied directly to the input data, namely: (i) the Okapi BM25 ranking function based approach described in [23]; and (ii) the n-grams approach, with a SVR ranking model, presented in [12] and [16]. Both were discussed previously in Subsect. 2.1.
3. To investigate the effect of the parameter k, the random walk length, on the operation of the GRAB-Rank approach.

The evaluation was conducted using a NVidia K80 GPUs kaggle kernel. The evaluation metrics used were precision and recall. Our dataset was labelled in a binary manner (relevant and not relevant) hence traditional document ranking metrics which require a "ground truth" ranking, such as MAP, MRR and NDCG, could not be used. For generating BERT embeddings all BERT Layers, as suggested in [21], were used. This was because using all layers produces a richer result, at the expense of increased run time. However, runtime is not an issue in the context of CDD maintenance as it is an activity not conducted frequently. In the case of the ORRCA CDD this is typically updated once every two years (because of the significant human resource involved). For the initial experiments conducted with respect to Objectives 1 and 2 above, $\sigma = 0.25$ was used with respect to the SVR LETOR models generated for reasons given previously in Sect. 5.

The results with respect to the first two objectives are given in Table 1. From the table it can be seen that in all cases the proposed GRAB-Rank hybrid approach produced the best performance with respect to both evaluation data sets and with respect to both recall and precision. From Table 1 it can also be seen that BERT only embedding tended to outperform knowledge graph embedding. The precision values are relatively low for the ORRCA-Update dataset. It is conjectured that this is because the ORRCA-Update dataset (25,584 documents) was significantly larger the ORRCA-400 dataset (400 documents).

To determine the most appropriate value for k experiments were conducted using a range of values for k from $k = 1$ to $k = 5$ incrementing in steps of 1. The results are presented in Table 2. From the table it can be seen that the best precision was obtained when $k = 2$ for the ORRCA-400 dataset, and $k = 3$ for the ORRCA-Update dataset. There was no clear best value for k with respect to

[2] https://www.nltk.org/.

Table 1. The performance of GRAB-Rank in comparison with using BERT embeddings or knowledge graph embeddings in isolation, and with using the BM25 and SVR ranking models with n-grams (best results in bold font).

Document ranking technique	ORRCA-400		ORRCA-Update	
	Precision	Recall	Precision	Recall
GRAB-Rank with SVR	**0.81**	**0.50**	**0.26**	**0.88**
BERT embeddings only with SVR	0.76	0.47	0.23	0.80
Knowledge graph embeddings only with SVR	0.75	0.46	0.26	0.87
word2vec vectors with BM25 ranking	0.53	0.33	0.16	0.54
word2vec for n-grams with SVR	0.79	0.49	0.07	0.49

recall. From the table it can also be seen that a low value of k ($k < 2$) produced poor recall. This could be attributed to the fact that the higher the value for k the more similar concepts that are included in the knowledge graph embedding and hence the better the recall (greater number of relevant documents at the top of a ranked document list).

Table 2. The performance of GRAB-Rank with using a range of values for k, the random walk length (best results in bold font).

k	ORRCA-400		ORRCA-Update	
	Precision	Recall	Precision	Recall
1	0.68	0.42	0.17	0.59
2	**0.75**	**0.46**	0.24	0.83
3	0.74	**0.46**	**0.26**	**0.87**
4	0.73	0.45	**0.26**	0.86
5	0.74	**0.46**	**0.26**	0.86

6   Future Work and Conclusion

This paper has presented the GRAB-Rank approach to partially automate the process of maintaining CDDs which would otherwise need to be maintained using a manual systematic review process. GRAB-Rank is a LETOR mechanism founded on a hybrid representation comprised of a literature knowledge graph embedding generated using a random walk of length k and a BERT contextual embedding. The hybrid embedding was then used as an input into a LETOR mechanism. For the presented evaluation SVR was used to generate the desired LETOR model. The hypothesis was that a hybrid document embedding approach would produce a better ranking than if the component embeddings were used in isolation. The GRAB-Rank approach was evaluated using two datasets extracted from the data used for the maintenance of the ORRCA CDD. The evaluation results obtained indicated that the hypothesis was correct, hybrid embedding outperforms individual embeddings used in isolation. The operation of GRAB-Rank was also compared with two forms of traditional approach, one based on BM25 and the other on n-gram based SVR. Grab-Rank was shown to outperform the traditional approaches. For future work the authors plan to improve the document ranking model by conducting further experiments, using the ORRCA CDD, with other pre-trained language model such as GPT-2 and GPT-3 for creating document embeddings.

References

1. Bagheri, E., Ensan, F., Al-Obeidat, F.: Neural word and entity embeddings for ad hoc retrieval. Inf. Process. Manag. **54**(4), 657–673 (2018)
2. Bollacker, K., Evans, C., Paritosh, P., Sturge, T., Taylor, J.: Freebase: a collaboratively created graph database for structuring human knowledge. In: Proceedings of the 2008 ACM SIGMOD International Conference on Management of Data, pp. 1247–1250 (2008)
3. Dai, Z., Xiong, C., Callan, J., Liu, Z.: Convolutional neural networks for soft-matching n-grams in ad-hoc search. In: Proceedings of the Eleventh ACM International Conference on Web Search and Data Mining, pp. 126–134 (2018)
4. Grover, A., Leskovec, J.: node2vec: scalable feature learning for networks. In: Proceedings of the 22nd ACM SIGKDD International Conference on Knowledge Discovery and Data Mining, pp. 855–864 (2016)
5. Guthrie, D., Allison, B., Liu, W., Guthrie, L., Wilks, Y.: A closer look at skip-gram modelling. In: LREC, vol. 6, pp. 1222–1225. Citeseer (2006)
6. Jabri, S., Dahbi, A., Gadi, T., Bassir, A.: Ranking of text documents using TF-IDF weighting and association rules mining. In: 2018 4th International Conference on Optimization and Applications (ICOA), pp. 1–6. IEEE (2018)
7. Kearney, A., et al.: Development of an online resource for recruitment research in clinical trials to organise and map current literature. Clin. Trials **15**(6), 533–542 (2018)
8. Li, Z., Guangluan, X., Liang, X., Li, F., Wang, L., Zhang, D.: Exploring the importance of entities in semantic ranking. Information **10**(2), 39 (2019)
9. Liu, Z., Xiong, C., Sun, M., Liu, Z.: Entity-duet neural ranking: understanding the role of knowledge graph semantics in neural information retrieval. arXiv preprint arXiv:1805.07591 (2018)
10. MacAvaney, S., Yates, A., Cohan, A., Goharian, N.: CEDR: contextualized embeddings for document ranking. In: Proceedings of the 42nd International ACM SIGIR Conference on Research and Development in Information Retrieval, pp. 1101–1104 (2019)
11. Mitra, M., Chaudhuri, B.B.: Information retrieval from documents: a survey. Inf. Retrieval **2**(2–3), 141–163 (2000). https://doi.org/10.1023/A:1009950525500
12. Muhammad, I., Bollegala, D., Coenen, F., Gamble, C., Kearney, A., Williamson, P.: Maintaining curated document databases using a learning to rank model: the ORRCA experience. In: Bramer, M., Ellis, R. (eds.) SGAI 2020. LNCS (LNAI), vol. 12498, pp. 345–357. Springer, Cham (2020). https://doi.org/10.1007/978-3-030-63799-6_26

13. Muhammad, I., Kearney, A., Gamble, C., Coenen, F., Williamson, P.: Open information extraction for knowledge graph construction. In: Kotsis, G., et al. (eds.) DEXA 2020. CCIS, vol. 1285, pp. 103–113. Springer, Cham (2020). https://doi.org/10.1007/978-3-030-59028-4_10
14. Nalisnick, E., Mitra, B., Craswell, N., Caruana, R.: Improving document ranking with dual word embeddings. In: Proceedings of the 25th International Conference Companion on World Wide Web, pp. 83–84 (2016)
15. Nogueira, R., Cho, K.: Passage re-ranking with BERT. arXiv preprint arXiv:1901.04085 (2019)
16. Norman, C.R., Gargon, E., Leeflang, M.M.G., Névéol, A., Williamson, P.R.: Evaluation of an automatic article selection method for timelier updates of the COMET Core Outcome set database. Database **2019**, 1–9 (2019). Article ID baz109. https://doi.org/10.1093/database/baz109
17. Padigela, H., Zamani, H., Croft, W.B.: Investigating the successes and failures of BERT for passage re-ranking. arXiv preprint arXiv:1905.01758 (2019)
18. Paik, J.H.: A novel TF-IDF weighting scheme for effective ranking. In: Proceedings of the 36th International ACM SIGIR Conference on Research and Development in Information Retrieval, pp. 343–352 (2013)
19. Pearson, K.: The problem of the random walk. Nature **72**(1867), 342–342 (1905)
20. Pennington, J., Socher, R., Manning, C.D.: GloVe: global vectors for word representation. In: Proceedings of the 2014 Conference on Empirical Methods in Natural Language Processing (EMNLP), pp. 1532–1543 (2014)
21. Peters, M.E., et al.: Deep contextualized word representations. arXiv preprint arXiv:1802.05365 (2018)
22. Rebele, T., Suchanek, F., Hoffart, J., Biega, J., Kuzey, E., Weikum, G.: YAGO: a multilingual knowledge base from Wikipedia, Wordnet, and Geonames. In: Groth, P., et al. (eds.) ISWC 2016. LNCS, vol. 9982, pp. 177–185. Springer, Cham (2016). https://doi.org/10.1007/978-3-319-46547-0_19
23. Shan, X., et al.: BISON: BM25-weighted self-attention framework for multi-fields document search. arXiv preprint arXiv:2007.05186 (2020)
24. Stanovsky, G., Michael, J., Zettlemoyer, L., Dagan, I.: Supervised open information extraction. In: Proceedings of the 2018 Conference of the North American Chapter of the Association for Computational Linguistics: Human Language Technologies, Volume 1 (Long Papers), pp. 885–895 (2018)
25. Uman, L.S.: Systematic reviews and meta-analyses. J. Can. Acad. Child Adolesc. Psychiatry **20**(1), 57 (2011)
26. Xiong, C., Dai, Z., Callan, J., Liu, Z., Power, R.: End-to-end neural ad-hoc ranking with kernel pooling. In: Proceedings of the 40th International ACM SIGIR Conference on Research and Development in Information Retrieval, pp. 55–64 (2017)
27. Xiong, C., Power, R., Callan, J.: Explicit semantic ranking for academic search via knowledge graph embedding. In: Proceedings of the 26th International Conference on World Wide Web, pp. 1271–1279 (2017)
28. Zamani, H., Croft, W.B.: Relevance-based word embedding. In: Proceedings of the 40th International ACM SIGIR Conference on Research and Development in Information Retrieval, pp. 505–514 (2017)

Advanced Analytics

Contextual and Behavior Factors Extraction from Pedestrian Encounter Scenes Using Deep Language Models

Jithesh Gugan Sreeram, Xiao Luo$^{(\boxtimes)}$ ⓘ, and Renran Tian ⓘ

Indiana University - Purdue University Indianapolis, Indianapolis, IN 46202, USA
jisree@iu.edu, {luo25,rtian}@iupui.edu

Abstract. This study introduces an NLP framework including deep language models to automate the contextual and behavior factors extraction from a narrative text that describes the environment and pedestrian behaviors at the pedestrian encounter scenes. The performance is compared against a baseline BiLSTM-CRF model trained for each factor separately. The evaluation results show that the proposed NLP framework outperforms the baseline model. We show that the proposed framework can successfully extract nested, overlapping, and flat factors from sentences through the case studies. This model can also be applied to other descriptions when physical context and human behaviors need to be extracted from the narrative content to understand the behavioral interaction between subjects further.

Keywords: Information extraction · Deep learning · Human behavior · Natural language processing · Autonomous vehicles

1 Introduction

According to [10], more than 6,500 pedestrians are killed in the year 2019, being the highest number in the last 30 years. The number is increased by more than 53% since 2009. Considering this upward trend, one of the primary purposes of deploying autonomous vehicles (AVs) is to reduce accidents and casualties to human pedestrians. The factors drawing human drivers' attention while interacting with the pedestrians are essential to adjust the importance of features while training the decision-making algorithms. One systematic way to study the reasoning process is through analyzing the drivers' descriptions of thoughts when assessing the pedestrian-vehicle encounters. Languages are the best-embodied knowledge of human intelligence. It is common to extract human factors during social or human-computer interactions by analyzing their verbal languages or written texts. In this research, we worked with human subjects who first watched a video recorded from the driver side and has a pedestrian encounter scene, then describe the context of the scene includes (1) The factors of the environment that influence the pedestrian-vehicle interaction, such as road structures, signal,

© Springer Nature Switzerland AG 2021
M. Golfarelli et al. (Eds.): DaWaK 2021, LNCS 12925, pp. 131–136, 2021.
https://doi.org/10.1007/978-3-030-86534-4_11

time of day, etc. (2) The features that dictate the interaction are between driver and pedestrian. (3) Pedestrian behaviors, such as walking patterns and so on. Figure 1 shows the one sentence in a description, in which 'looked at driver and started running' is an implicit communication between driver and pedestrian, 'started running across the crosswalk' is a pedestrian walking pattern, 'crosswalk' is a road structure, and 'signal is green' is a signal. Both 'crosswalk' and 'signal is green' are contextual factors of the environment. All these are important to understand the scene and advise the future development and deployment of autonomous vehicles. Our objective is to develop an NLP framework to extract these factors from pedestrian encounter descriptions in this research.

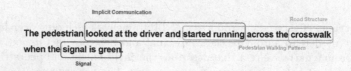

Fig. 1. A sentence in the pedestrian encounter description (Color figure online)

Named Entity Recognition (NER) has been used to identify and extract concepts from narrative text in various domains [1,4]. The recent neural network-based NER models show superior performances to the traditional ones. However, with a typical NER model, each word in the sentence is predicted to be part of a concept of one category. For example, given a sentence 'The pedestrian is walking slowly towards the traffic light.', 'walking slowly' is pedestrian speed, and 'traffic light' is signal. They belong to two different categories, and there is no overlapping between these two concepts. NER models can be trained to extract both concepts. However, if the concepts belong to multiple categories are nested or overlapped, shown as Fig. 1, the typical NER model has limitations on extracting all concepts correctly. The reason is that the number of categories for each word is different, and there can be multiple BIOE tags for each word corresponding to different categories. In this research, we call it multiclass multilabel concept extraction. Recently, NER models worked with pre-trained contextual embeddings such as Word2Vec [7], Glove [8] or BERT [2]. Fisher et al. [3] introduced a model first to identify the boundaries of the named entities and the nesting situation at different levels using the BERT embedding. This approach is not applicable to the overlapping entities. Shibuya et al. [11] proposed a sequence decoding algorithm with a CRF model to extract the nested entities. Strakova et al. [12] compared a sequence-to-sequence model with LSTM-CRF for nested entity recognition and concluded that the sequence-to-sequence approach worked better. However, multiple sequence-to-sequence models need to be trained for each category to solve the multilabel multiclass issue.

To solve multilabel and multiclass concept extraction challenges, we propose an NLP framework explained in Sect. 2. The returned results show that our approach works better than the baseline approach.

2   Methodology

Figure 2 shows an overview of the proposed framework. First, a NER model –
BiLSTM-CRF [5,6] is used to extract text chunks that contain one or more
concepts. BiLSTM-CRF takes word embeddings and tags the text chunks using
the BIOE scheme (B-, I-, E-represents the start, internal, and end of a text
chunk, and O- tags the word outside of a text chunk).

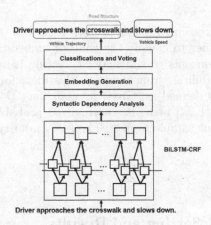

Fig. 2. Contextual and behaviour factors extraction framework

Then, syntactic dependencies are analyzed to obtain the possible concepts
from the text chunks. Syntactic dependency analysis [13] is used to extract candidate concepts from the text chunks identified by BiLSTM-CRF. In our approach,
we generate candidate concepts with a noun or a verb as a centered word. Taken
a noun or verb as a target word (w_i) in a sentence, we first identify all the direct
and indirect incoming and outgoing edges (E) from the word, then we extract
all the candidate concepts based on the edge connections to the target words.
For example, if the word 'pace' is taken as a target word, E are edges shown
in Fig. 3. A candidate concept is defined as $c = \{w_0, \ldots, w_i, \ldots, w_j\}$, and edge
$e = (w_i, w_j)$, $e \in E$ between any two words in the c. Based on the words and
edges in Fig. 3 and the following candidate concepts can be extracted: 'pace',
'fast pace', 'at fast pace', 'walking at a fast pace', and 'begins walking at fast
pace'.

After concept extraction, a transformer-based embedding generation is
employed to generate vector representations of all concepts. The embedding
techniques are used in various research to generate word or phrase representations. In this research, the sentence transformer (SBERT) based on RoBERTa
[9] is used to generate embeddings for the candidate concepts.

Finally, the vector representations of the concepts are fed into different classifiers for concept classification and prediction. We implement four different

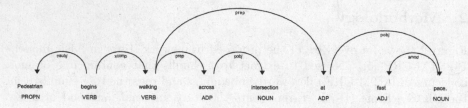

Fig. 3. An sentence example for syntactic analysis

classifiers: logistic regression, random forest, K-nearest neighbor, and multi-layer perceptron classifier to classify the concept embeddings into corresponding categories of the concepts. A soft majority voting is implemented to aggregate the output probabilities from different classifiers for the categories. The soft majority voting is defined as: Given M classifiers $D = \{D_1, \ldots, D_M\}$, let $y_j(x_i)$, $i \in [1, N]; j \in [1, M]$ being the normalized probability output from the classifier j, i is an input sample from x. The soft majority voting is calculated as Eq. 1.

$$\sum_{j=1}^{M} y_j(x_i) \tag{1}$$

3 Experimental Setting and Results

3.1 Dataset and Hyperparameters

In this research, we recruited 30 human subjects to describe 100 different pedestrian encounter scenes. To develop the labeled dataset, two annotators worked separately to annotate the sentences in the descriptions to identify the contextual and behavior factors belongs to 13 different categories given in Table 1. In total, we have 1005 sentences to evaluate our concept extraction framework. The overall data set is split into train, test, and validation, with 70%, 20%, and 10% of the sentences, respectively. For both the baseline and our model, 100 neurons are set for BiLSTM, and ten epochs are used for training. For the random forest classifier, the number of trees is set to 100. The k is set to 5 for the k-nearest neighbor classifier. For logistic regression, multiclass is set to multinomial. For the multilayer perceptron classifier, there is only one hidden layer with 100 neurons.

3.2 Results

The BiLSTM-CRF is trained for each category as the compared baseline method. The precision, recall, and F1 results are calculated using the test set for both the baseline model and our model, shown in Table 1. Our framework works better than the baseline method on most categories, including those with very few instances, such as 'lighting condition', and 'scene location'. Our model also works better on the categories that have nested or partial overlapping content.

Table 1. Comparison of results between baseline and our approach on test data

Category	Number	BiLSTM-CRF			Our model		
		Precision	Recall	F1	Precision	Recall	F1
Pedestrian walking pattern	218	0.71	0.63	0.66	**0.75**	**0.85**	**0.8**
Pedestrian speed	92	0.72	0.63	0.67	0.27	0.32	0.29
Road structure	169	0.25	0.33	0.29	**0.96**	**1**	**0.98**
Time of day	17	0	0	0	**0.86**	**0.86**	**0.86**
Signal	61	1	0.1	0.18	**0.97**	**0.93**	**0.95**
Vehicle trajectory	117	0.55	0.34	0.42	0.51	**0.87**	**0.64**
Vehicle speed	61	0.55	0.34	0.42	**0.95**	0.42	**0.58**
Lighting condition	14	0	0	0	**0.5**	**0.5**	**0.5**
Scene location	30	1	0.2	0.33	1	**0.71**	**0.83**
Vehicle type	79	0.83	0.69	0.76	**1**	**1**	**1**
Pedestrian attention	30	0.5	0.09	0.15	**1**	**0.36**	**0.53**
Explicit communication	33	1	0.17	0.29	0.77	**0.59**	**0.67**
Implicit communication	201	0.55	0.27	0.36	**0.56**	**0.88**	**0.68**

Pedestrian begins
walking across
intersection at fast pace.

Fig. 4. Case 1 - baseline model

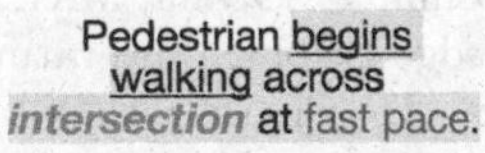

Pedestrian begins
walking across
intersection at fast pace.

Fig. 5. Case 1 - our model

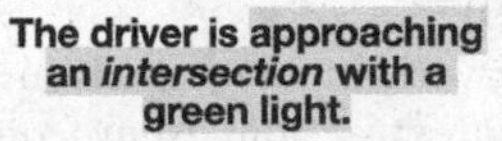

The driver is approaching
an *intersection* with a
green light.

Fig. 6. Case 2 - baseline model (Color figure online)

The driver is approaching
an *intersection* with a
green light.

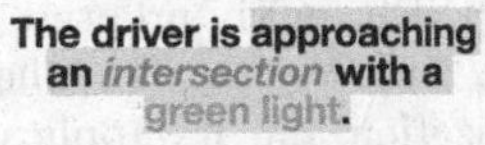

Fig. 7. Case 2 - our model (Color figure online)

Figures 4 and 5 show a sentence with concepts of four different categories. Our model can correctly identify all concepts to the corresponding categories. 'begins walking', 'intersection', 'fast pace' and 'begins walking across intersection at a fast pace' are identified as implicit communication, road structure, pedestrian speed, and pedestrian walking pattern, respectively. The baseline model can only identify 'begins walking across intersection at fast pace' as pedestrian walking pattern and 'intersection' as road structure. Figures 6 shows that the baseline model can extract 'approaching an intersection with a green light' as 'vehicle trajectory', but it can not identify other concepts nested in this long description. As shown in Fig. 7, our approach not only can extract 'approaching an intersection with a green light' as vehicle trajectory but also extracts 'intersection' as road structure and 'green light' as a signal. This case also shows that our approach can successfully extract all the nested concepts from a long description without sacrificing the accuracy of any categories.

4   Conclusions and Future Work

This research investigates an NLP framework to extract contextual and behavior factors from descriptions about pedestrian encounters in traffic scenes. The concepts that describe the factors can be nested, overlapping, or flat in the same sentence. The objective is different from the traditional NER tasks, which generally work with name entities. The results show that our model works better than the baseline model on most of the categories, especially those nested within or overlapping with other categories. The limitation of our model is that when two concepts that belong to multiple categories are the same with different targets, such as vehicle speed and pedestrian speed, our model cannot differentiate them properly. The future work is to enhance the system to analyze the context around those concepts through integrating syntactic and semantic analysis.

References

1. Al-Moslmi, T., Ocaña, M.G., Opdahl, A.L., Veres, C.: Named entity extraction for knowledge graphs: a literature overview. IEEE Access **8**, 32862–32881 (2020)
2. Devlin, J., Chang, M.W., Lee, K., Toutanova, K.: BERT: pre-training of deep bidirectional transformers for language understanding. arXiv preprint arXiv:1810.04805 (2018)
3. Fisher, J., Vlachos, A.: Merge and label: a novel neural network architecture for nested NER. arXiv preprint arXiv:1907.00464 (2019)
4. Li, L., Xu, W., Yu, H.: Character-level neural network model based on Nadam optimization and its application in clinical concept extraction. Neurocomputing **414**, 182–190 (2020)
5. Lin, B.Y., Xu, F.F., Luo, Z., Zhu, K.: Multi-channel BILSTM-CRF model for emerging named entity recognition in social media. In: Proceedings of the 3rd Workshop on Noisy User-generated Text, pp. 160–165 (2017)
6. Lopes, F., Teixeira, C., Oliveira, H.G.: Comparing different methods for named entity recognition in Portuguese neurology text. J. Med. Syst. **44**(4), 1–20 (2020)
7. Mikolov, T., Chen, K., Corrado, G., Dean, J.: Efficient estimation of word representations in vector space. arXiv preprint arXiv:1301.3781 (2013)
8. Pennington, J., Socher, R., Manning, C.D.: Glove: global vectors for word representation. In: Proceedings of the 2014 Conference on Empirical Methods in Natural Language Processing (EMNLP), pp. 1532–1543 (2014)
9. Reimers, N., Gurevych, I.: Sentence-BERT: sentence embeddings using Siamese BERT-networks. In: Proceedings of the 2019 Conference on Empirical Methods in Natural Language Processing. Association for Computational Linguistics (11 2019). https://arxiv.org/abs/1908.10084
10. Retting, R.: Pedestrian traffic fatalities by state: 2019 preliminary data (2020)
11. Shibuya, T., Hovy, E.: Nested named entity recognition via second-best sequence learning and decoding. arXiv preprint arXiv:1909.02250 (2019)
12. Straková, J., Straka, M., Hajič, J.: Neural architectures for nested NER through linearization. arXiv preprint arXiv:1908.06926 (2019)
13. Vasiliev, Y.: Natural Language Processing with Python and SpaCy: A Practical Introduction. No Starch Press (2020)

Spark Based Text Clustering Method Using Hashing

Mohamed Aymen Ben HajKacem[1]([✉]), Chiheb-Eddine Ben N'Cir[1,2], and Nadia Essoussi[1]

[1] LARODEC, Institut Supérieur de Gestion de Tunis,
Université de Tunis, Tunis, Tunisia
{chiheb.benncir,nadia.essoussi}@isg.rnu.tn
[2] College of Business, University of Jeddah, Jeddah, Saudi Arabia

Abstract. Text clustering has become an important task in machine learning since several micro-blogging platforms such as Twitter require efficient clustering methods to discover topics. In this context, we propose in this paper a new Spark based text clustering method using hashing. The proposed method deals simultaneously with the issue of clustering huge amount of documents and the issue of high dimensionality of textual data by respectively fitting parallel clustering of large textual data through Spark framework and implementing a new document hashing strategy. Experiments performed on several large collections of documents have shown the effectiveness of the proposed method compared to existing ones in terms of running time and clustering accuracy.

Keywords: Textual data · K-means · Spark · Hashing

1 Introduction

Text clustering also known as document clustering, aims to group text documents, described by set of terms, into clusters so that documents within a cluster discuss similar topics. Several clustering methods were proposed in the literature [11] which can be categorized into hierarchical, density-based, grid-based methods, model-based and partitional [7]. Among these categories, hierarchical and partitional categories are the widely used in text mining applications because of their efficiency [9].

Although the attested performance of existing text clustering methods, they suffer from the scalability issue. Given the continuous growth of large documents, designing a scalable text clustering method becomes an important challenge since conventional clustering methods cannot process such large amount of documents. For example, k-means clustering method, does not scale with huge volume of documents. This is explained by the high computational cost of these methods which require unrealistic time to build the grouping.

To deal with large scale data, several clustering methods which are based on parallel frameworks have been designed in the literature [2–4,16]. Most of

© Springer Nature Switzerland AG 2021
M. Golfarelli et al. (Eds.): DaWaK 2021, LNCS 12925, pp. 137–142, 2021.
https://doi.org/10.1007/978-3-030-86534-4_12

these methods use the MapReduce to distribute computing across a cluster of machines. For instance, Zaho el al. [16] proposed fitting k-means method through MapReduce framework [5]. This method first assigns each data point to the nearest cluster center in the map phase and updates the new cluster centers in the reduce phase. Then, this method iterates calling the two phases several times until convergence. Cui et al. [4] also proposed an optimized MapReduce-based k-means method, which is based on generating data sample from data using probability sampling to improve the efficiency of clustering when dealing with large scale data. Another parallel method was also proposed by Ben HajKacem et al. [2], which is based on a pruning strategy to reduce the number comparisons between data points and cluster centers.

Despite existing parallel methods providing an efficient analysis of large scale data using MapReduce, they are not explicitly designed to deal with huge amount of textual data. In addition, MapReduce framework is unsuitable to run iterative algorithms since it requires at each iteration reading and writing data from disks. On the other hand, the application of these methods to text clustering needs to prepare the collection of documents for a numerical analysis by using text representation method such as the vector space model (VSM) [10] which is one of the widely used text representation method in information retrieval and text mining applications [8]. In fact, the construction of this representation is time consuming especially when dealing with large documents. Furthermore, this representation suffers from the high-dimensionality resulted feature space, which makes calculations in the vector space computationally expensive.

To deal with all these issues, we propose in this paper a new Spark based Text Clustering Hashing, referred to as STCH. The proposed method is based on distributing data across several machines and parallelizing the text clustering process to be executed on each local machine using Spark framework [15] which has showed faster execution of parallel tasks compared to MapReduce [13]. Furthermore, to improve the efficiency of parallel text clustering tasks, we propose a new document hashing strategy which is based on reducing the dimensionality of feature space using the hashing trick technique [14].

The remainder of this paper is organized as follows: Sect. 2 presents the hashing trick technique. After that, Sect. 4 describes the proposed STCH method while Sect. 5 presents experiments that we have performed to evaluate the efficiency of the proposed method. Finally, Sect. 7 presents conclusion and future works.

2   The Hashing Trick

The hashing trick was first presented in [14] and then extended in [1]. Given X a representation of data in a t-dimensional feature space, indexed by $[t]=\{1\ldots t\}$ and let h:$[t] \rightarrow U=\{1\ldots u\}$ with $u \ll t$ be a hash function drawn from a family of pairwise independent hash functions. Given $x \in X$ a vector, the hashing trick maps the original representation x to a new hashed representation $\phi(x)$ in the new low dimensional space indexed by U, where the elements of $\phi(x)$ are defined as follows:

$$\phi(x)_i = \sum_{j:h(j)=i} x_j, \tag{1}$$

3 Spark Based Text Clustering Method for Large Textual Data

The proposed method deals with the issues of clustering huge amount of documents and also with the high dimensionality of textual data by respectively distributing clustering process through Spark framework and implementing a new document hashing strategy.

STCH method consists of two MapReduce jobs namely; *Document hashing* and *Document clustering*. The document hashing MapReduce job is devoted to build a hashed low-dimensional feature space of documents from a set of documents. The document clustering MapReduce job consists of clustering the new feature space to look for documents representatives.

3.1   Document Hashing MapReduce Job

This job aims to generate an approximate low dimensional feature space rather than the high original dimensional space of the set of documents. First, documents are divided into set of m chunks where each chunk is mapped to a low dimensional feature space using the hashing trick technique. Then, the m intermediate dimensional feature spaces are collected and grouped in the reduce phase. In order to explain this MapReduce job, we first explain how we performed the document hashing and then, we explain how we implemented this job through Spark framework.

Document Hashing Strategy: We present in this section the proposed distributed strategy to approximate the TF representation using hashing. An important advantage of this strategy is its ability to compute the hashed representation by simply scanning the words in d one by one, without building vocabulary and without generating the TF representation (VSM construction). Note that the hashing function h:W $\rightarrow$ U $= \{1, 2 \ldots u\}$ operates directly on words by summing the ASCII values of the letters in given word. Hence, each document d is hashed to $\phi(d)$ by summing ASCII values of letters in each word in this document. The element $\phi(d)_g$ of the new hashed representation is defined as follows:

$$\phi(d)_g = \sum_{w:h(w)=g} TF(w, d), \tag{2}$$

where g is the index assigned to the word w using hash function h and $TF(w, d)$ is the number of occurrences of the word w in document d.

Parallel Implementation: We present in this section the implementation of the document hashing strategy through Spark framework. Given that each document has its independent hashed representation, we propose to execute these hashing processes in parallel by implementing a MapReduce job and using the distributed Spark framework. First, an RDD object with m chunks is created from input documents D. In the following we detail the map and reduce phases.

- **Map phase**: During this phase, each map task picks a chunk, computes the hashed TF representation for each document and generates the intermediate hashed feature spaces from this chunk. Then, it emits the intermediate feature space as the output.
- **Reduce phase**: Once the map phase is finished, a set of intermediate feature spaces is emitted to a single reduce phase. The reduce phase collects the set of intermediate feature spaces and returns the final feature space of input document dataset as the output.

3.2 Document Clustering MapReduce Job

Once the resulted feature space is generated, the next step is to built the distributed partitioning of documents in a separate MapReduce job containing map and reduce phases. First, an RDD object with m chunks is created from the resulted hash space Φ. In the following we detail the map and reduce phase.

- **Map phase**: During this phase, the map function takes a chunk, executes k-means on this chunk and extracts the cluster documents representatives. Then, the map function emits the extracted intermediate cluster documents representatives as the output to a single reduce phase.

- **Reduce phase**: The reduce phase takes the set of intermediate cluster documents representatives, executes again the k-means and returns the final cluster documents representatives as the output.

4   Experiments and Results

4.1   Methodology

In order to evaluate the performance of STCH method, we performed experiments on real large text datasets. During the evaluation, we tried to figure out this point: How much is the efficiency and the speed of STCH when applied to collections of large documents compared to existing methods namely, MapReduce based k-means (MRKM) [16] and Min-batch k-means (MBKM) [12]?

4.2   Environment and Datasets

The experiments are performed on a cluster of 8 machines where each machine has 1-core 2.30 GHz CPU E5400 and 1GB of memory. The experiments are realized using Apache Spark version 1.6.2, Apache Hadoop 1.6.0 and Ubuntu 14.04. We conducted experiments on Hitech, 20newsgroups (20newg) and Reuters datasets which are summarized in Table 1.

Table 1. Summary of the document datasets

Dataset	Number of documents	Number of terms	Number of classes
Hitech	2.301	13.170	6
20newg	4.345	54.495	20
R52	6.532	15.596	52
R8	5.485	14.009	8

4.3   Results

In this section, we first evaluate the running time of STCH with existing methods based on VSM representation. The results are reported in Table 2. These results show that STCH is always faster than existing methods for all datasets. For example, Table 2 shows that STCH method is 5 times faster than MRKM for R8 dataset. Then, we evaluate the clustering quality of the proposed method compared to existing ones. We used external validation measures namely Precision, Recall and F-measure. The obtained results show that STCH produces a comparable accuracy compared to existing methods for all datasets.

Table 2. Comparison of the results of STCH with existing methods on real datasets

Dataset	Method	Running time (seconds)	Precision	Recall	F-measure
R8	MRKM	66.27	0.834	0.565	0.568
	MBKM	82.27	0.188	0.180	0.158
	STCH	**13.08**	0.755	0.441	0.481
R52	MRKM	70.27	0.834	0.565	0.568
	MBKM	83.77	0.188	0.328	0.196
	STCH	**12.53**	0.817	0.368	0.366
Hitech	MRKM	77.17	0.834	0.565	0.568
	MBKM	84.94	0.188	0.328	0.196
	STCH	**18.12**	0.249	0.242	0.224
20newg	MRKM	704.27	0.834	0.565	0.568
	MBKM	87.09	0.194	0.184	0.182
	STCH	**35.27**	0.656	0.339	0.423

5   Conclusion

In order to deal with the issues of identifying clusters of text documents and also with the high dimensionality of textual data, we have proposed a new parallel text clustering method which is based on fitting text clustering process through Spark framework and implementing a new document hashing strategy. The obtained results on large real text datasets show the efficiency of the proposed method compared to existing ones using VSM. The application of STCH to large text documents gives a promising result which could be improved by integrating external knowledge such as Wordnet or Wikipedia. In addition, the STCH could be improved by quantifying the relatedness of a term to a cluster by its discrimination information.

References

1. Attenberg, J., Weinberger, K., Dasgupta, A., Smola, A., Zinkevich, M.: Collaborative email-spam filtering with the hashing trick. In: The Sixth Conference on Email and Anti-Spam (2009)
2. Ben HajKacem, M.A., Ben N'cir, C.-E., Essoussi, N.: One-pass MapReduce-based clustering method for mixed large scale data. J. Intell. Inf. Syst. **52**(3), 619–636 (2017). https://doi.org/10.1007/s10844-017-0472-5
3. Ben HajKacem, M.A., Ben N'Cir, C.-E., Essoussi, N.: Overview of scalable partitional methods for big data clustering. In: Nasraoui, O., Ben N'Cir, C.-E. (eds.) Clustering Methods for Big Data Analytics. USL, pp. 1–23. Springer, Cham (2019). https://doi.org/10.1007/978-3-319-97864-2_1
4. Cui, X., Zhu, P., Yang, X., Li, K., Ji, C.: Optimized big data K-means clustering using MapReduce. J. Supercomputing **70**(3), 1249–1259 (2014)
5. Dean, J., Ghemawat, S.: MapReduce: simplified data processing on large clusters. Commun. ACM **51**(1), 107–113 (2008)
6. Fraj, M., Ben Hajkacem, M.A., Essoussi, N.: On the use of ensemble method for multi view textual data. J. Inf. Telecommun. **4**(4), 461–481, 81–99 (2020)
7. Jain, A.K., Murty, M.N., Flynn, P.J.: Data clustering: a review. ACM Comput. Surv. (CSUR) **31**(3), 264–323 (1999)
8. Lan, M., Tan, C.L., Su, J., Lu, Y.: Supervised and traditional term weighting methods for automatic text categorization. IEEE Trans. Pattern Anal. Mach. Intell. **31**(4), 721–735 (2009)
9. Landset, S., Khoshgoftaar, T.M., Richter, A.N., Hasanin, T.: A survey of open source tools for machine learning with big data in the Hadoop ecosystem. J. Big Data **2**(1), 1–36 (2015). https://doi.org/10.1186/s40537-015-0032-1
10. Salton, G.: Automatic Text Processing: The Transformation, Analysis, and Retrieval of Information by Computer. Addison-Wesley, Reading (1989)
11. Saxena, A., et al.: A review of clustering techniques and developments. Neurocomputing **267**(2017), 664–681 (2017)
12. Sculley, D.: Web-scale k-means clustering. In: The 19th International Conference on World Wide Web, ACM, pp. 1177–1178 (2010)
13. Singh, D., Reddy, C.K.: A survey on platforms for big data analytics. J. Big Data **2**(1), 1–20 (2014). https://doi.org/10.1186/s40537-014-0008-6
14. Shi, Q., Petterson, J., Dror, G., Langford, J., Smola, A., Vishwanathan, S.V.N.: Hash kernels for structured data. J. Mach. Learn. Res. **10**(2009), 2615–2637 (2009)
15. Zaharia, M., Chowdhury, M., Franklin, M.J., Shenker, S., Stoica, I.: Spark: cluster computing with working sets. HotCloud **10**(10–10), 95 (2010)
16. Zhao, W., Ma, H., He, Q.: Parallel k-means clustering based on mapreduce. In: The IEEE International Conference on Cloud Computing, pp. 674–679. Springer (2009)

Impact of Textual Data Augmentation on Linguistic Pattern Extraction to Improve the Idiomaticity of Extractive Summaries

Abdelghani Laifa[1]([⊠])(iD), Laurent Gautier[2,3]([⊠])(iD), and Christophe Cruz[3]([⊠])(iD)

[1] Laboratoire d'Informatique de Bourgogne, Maison des Sciences de l'Homme, Université de Bourgogne, Dijon, France
Abdelghani_Laifa@etu.u-bourgogne.fr
[2] Maison des Sciences de l'Homme, Université de Bourgogne, Dijon, France
Laurent.Gautier@u-bourgogne.fr
[3] Laboratoire d'Informatique de Bourgogne LIB, EA 7534, Univ. Bourgogne Franche-Comté, Dijon, France
Christophe.cruz@u-bourgogne.fr

Abstract. The present work aims to develop a text summarisation system for financial texts with a focus on the fluidity of the target language. Linguistic analysis shows that the process of writing summaries should take into account not only terminological and collocational extraction, but also a range of linguistic material referred to here as the "support lexicon", that plays an important role in the cognitive organisation of the field. On this basis, this paper highlights the relevance of pre-training the CamemBERT model on a French financial dataset to extend its domain-specific vocabulary and fine-tuning it on extractive summarisation. We then evaluate the impact of textual data augmentation, improving the performance of our extractive text summarisation model by up to 6%–11%.

Keywords: Text summarisation · Terminology · Linguistic patterns · Natural language processing · Corpus linguistics · Deep learning

1 Context and Objectives

The work presented here is part of a larger project conducted at the University of Burgundy, at the crossroads between the language sciences and data science. The main research question is how to extract patterns from their environment in order to improve the readability of automatic summaries within the field of finance. The project also questions, as in the following case study around what we propose to call the "support lexicon", the limits of strictly terminological and/or collocational inputs when dealing with specialised discourses. We explore a set of economic reports from the *Banque de France*, which are "serial texts" that combine various statistics with more explicative sequences. The extraction based on the usual approaches in terminology quickly showed that an essential

M. Golfarelli et al. (Eds.): DaWaK 2021, LNCS 12925, pp. 143–151, 2021.
https://doi.org/10.1007/978-3-030-86534-4_13

part of the lexicon was neglected. We therefore considered this to be a limit of such approaches and explored the possibilities of focusing on the support lexicon and the information it conveys, especially for deep learning. Thus, we present an extractive summarisation model for French based on self-attention. In this paper, we highlight our method of extracting lexico-grammatical patterns using the attention mechanism, which is a part of a neural architecture capable of highlighting relevant features in the input data. Because of the limited number of monthly reports available, we used the data augmentation principle to generate artificial data from the original dataset. The augmented corpus, composed of 226 initial reports and 226 artificial reports, allowed us to improve the performance of the fine-tuned model.

The corpus belongs to a discourse type that can be qualified as "conjuncture discourse" [1, 6, 20] and is produced by national central banks. It generally presents a very high degree of recurrence: both at the level of the contents themselves and of the form, with a very rigid macrostructure. The corpus is in French, and each report has its own summary. From a quantitative perspective, the corpus contains 323 monthly reports published between 1994 and 2020, and is made up of 6,554,396 words, 4,070,955 lemmas and 317,076 sentences. Each report has an average length of 1500 words, and each summary has an average length of 200 words. The remainder of this paper is structured as follows. Section 2 provides elements about the limits of terminology-based models and introduces automatic text summarisation. Section 3 presents the approach adopted here to produce summaries, starting with an explanation of the pre-training and the fine-tuning processes, as well as the extraction of syntactic and semantic patterns. It ends with the data augmentation method. Section 4 describes the results and evaluations.

2   Background

Pattern extraction based solely on the terminological approach has been revealed to be limited. The extraction of lexico-grammatical patterns based on attention mechanisms therefore seemed to be a solution to improve the idiomaticity of the summaries produced.

2.1   From Terminology to Patterns and Constructions

Recent work on specialised discourses has very clearly emphasised the limits of approaches based on words as isolated units. The continuous extension of the field of "phraseology" [5, 8, 13, 14] aims to capture recurrent segments that are more significant than words. Three holistic approaches are currently implemented in this field, revealing the role played by support lexicon, as a lexicon which does not fall within traditional terminology but is "applied", through patterns, to other terminology-collocational structures [12].

Frame Semantics already implemented in terminology, is a cognitive linguistics paradigm aimed at an organised representation of knowledge linked to a concept that results from the experience of the speaker. Frames lead to particular encodings in language through combinatorics and preferential syntactic productions. **Lexico-grammatical patterns** challenge the traditional view of lexical modules (dictionaries), on which grammatical rules operate. Not only does the type of constrained text studied here implement a restricted repertoire of the French grammatical language system, but it does so only in synergy with the lexicon concerned with building blocks of meaning [7]: "The typical linguistic features of ESP cannot be characterised as a list of discreet items (technical terminology, the passive, hedging, impersonal expressions, etc.), rather the most typical features of ESP texts are chains of meaningful interlocking lexical and grammatical structures, which we have called lexico-grammatical patterns". They are the markers of a double idiomaticity in the corpus: idiomaticity of the language, and idiomaticity of the field allowing the experts to convey field-related information without ambiguity. **Construction grammars** represent the highest degree of abstraction and generalisation of frames and allow the syntax-semantic interface to be modeled with a high degree of granularity. As they are usage-based, they also have their starting point in recurring patterns.

2.2 Text Summarisation

Text summarisation consists of creating a short version of a text document by extracting the essential information. There are two main approaches: extractive and abstractive. The extractive approach extracts the document's most salient sentences and combines them into a summary. In contrast, the abstractive approach aims to generate a summary as humans do, by extracting and paraphrasing the original text. There are relatively few works on text summarisation in French compared to other languages, and they focus mainly on extractive approaches by scoring sentences and selecting the highly scored ones for the summary. The lack of a French benchmark corpus makes evaluation for French summarisation more difficult. CamemBERT [17] is the first BERT-type [2] language model. It was trained on a French dataset containing the texts of numerous web pages.

3   Methodology

Fundamentally, the BERT-type model is a stack of Transformer encoder layers [23] that consist of multiple self-attention "heads". For every input token in a sequence, each head computes key, value, and query vectors, used to create a weighted representation. The outputs of all heads in the same layer are combined and run through a fully connected layer. Each layer is wrapped with a skip connection and followed by layer normalisation. The input representations are computed as follows: Each word in the input is first tokenised with SentencePiece [11], and then three embedding layers (token, position, and segment) are combined to obtain a fixed-length vector. Special token [CLS] is used for classification predictions because it stores the combined weights provided by the heads of each sentence, and [SEP] separates the input segments. We will break down our approach into two essential stages, pre-training and fine-tuning. We first continued the training of the original CamemBERT model on the financial dataset using the masked language modeling task. Then, we fine-tuned our pre-trained CamemBERT model on an extractive summarisation task.

3.1   Pre-training the Original CamemBERT on a Financial Dataset

In this method, we inserted all the corpus texts except the summaries into the model. The model learned the domain-specific financial vocabulary as well as the reports' lexico-grammatical patterns by predicting the randomly masked input tokens. The idea behind this method is to expand the knowledge of the original CamemBERT and to create a new French CamemBERT model that can understand and evaluate textual data in the field of finance.

3.2   Fine-Tuning of Our Model on Extractive Summarisation

The fine-tuning method consists of adding a sentence classifier on top of the final encoder layer to predict which sentences should be included in the summary. For this, we converted our summarisation dataset, which consists of report-summary peers, to an extractive summarisation dataset by assigning a label 1 to each sentence of the report included in the summary and a label 0 to the other sentences in the report. Then, we divided the corpus into a training (226 reports), a validation (48 reports) and a test dataset (49 reports), in order to train and test the sentence classifier. At the end of this process, we froze our model so that only the parameters of the sentence classifier were learned from scratch. The model was pre-trained for 2 epochs over 44,800 steps on 1 GPU (Tesla K80) with a learning rate of $5e^{-5}$, setting a batch size for training to 10 and 100 for fine-tuning. Model checkpoints are saved and evaluated on the validation set every 500 steps. The sentence classifier was fine-tuned on the same GPU for 3 epochs with a learning rate of $2e^{-5}$.

3.3   Heads with Linguistic Knowledge

CamemBERT uses 12 layers of attention, and also incorporates 12 attention "heads" in every layer. Since model weights are not shared between layers, the CamemBERT model has up to $12 \times 12 = 144$ different heads of attention mechanisms. The question is: what is the capability of the encoder's attention mechanism in capturing linguistic knowledge as lexico-grammatical patterns? We extracted the attention mapped from our pre-trained model with BertViz [24] to explore the attention patterns of various layers/heads and to determine the linguistic patterns that mapped to our understanding of lexico-grammatical parsing.

Syntactic Dependency. After extracting the attention maps, we evaluated the prediction direction for each attention head: the direction of the head's word referring to the dependent and the opposite one. Three syntactic dependencies were explored to emphasise the interdependence of vocabulary (lexis) and syntax (grammar). **Results:** As Fig. 1 shows, in (A), heads 4 and 12 of layer 6 allowed the verbs "redresser" and "inscrire" to correctly address their attention to their subject. In (B), heads 4 and 11 of layer 4 allowed the noun's determiner to attend their noun, and in (C) the head 2 of layer 1 allowed the prepositions to attend their complements. Our model's attention heads performed remarkably well, illustrating how syntactic dependencies can emerge from the pre-training phase. We also noticed that other attention heads were able to capture these syntactic patterns but with less significant attention. We also experimented with the attention heads of the original CamemBERT on the financial data. The results were not significant because the attention weights were dispersed over the whole sequence without taking into account the grammar of the specific domain, in particular for experiment (A).

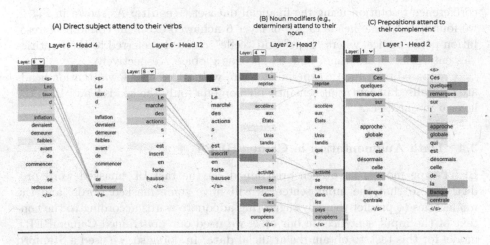

Fig. 1. CamemBERT attention heads that correspond to some syntactic linguistic phenomena. In this visualisation, the darkness of a line indicates the strength of the attention weight. All attention to/from words is coloured according to the head's colour, which emphasises the attention pattern. Selected words are highlighted with a grey box.

Semantic Dependency. Coreference resolution is a challenging task that requires context understanding, reasoning, and domain-specific perception. It consists of finding all the linguistic expressions in a given text that refer to the same real world entity. This is why most of the linguistic phenomena studies [10,16,25] have been devoted to BERT's syntactic phenomena rather than its semantic phenomena [3,22]. We used attention heads for the challenging semantic task of coreference resolution. We evaluated the attention heads of our model on

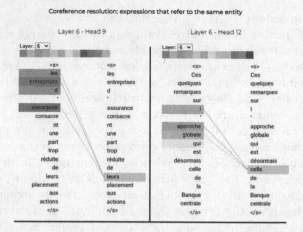

Fig. 2. Visualisation of coreference resolution attentions.

coreference resolution using the financial dataset. **Results:** As shown in Fig. 2, we found that some heads (9–12) of layer 6 achieved excellent coreference resolution performance, where "leurs" and "celle" correctly referred to their entities "les enterprises d'assuranc" and "l'approche globale" respectively.

Compared to syntactic experimentation, we noticed that few attention heads are specialised in capturing semantic phenomena (only the heads 9 and 12 of the layer 6 in our case).

3.4   Data Augmentation in CamemBERT

BERT-type models have been pre-trained on the task of "masked word prediction", so that the input sentences will have some masked words, and the model tries to predict them by suggesting adequate words according to the context of the input sentence. In our case, we used our pre-trained CamemBERT model for this task to obtain the artificial data. In doing so, we used a Stanford part-of-speech tagger trained in French to read the dataset reports and assign parts of speech to each word. Then, we selected all the modifiers (adjectives and adverbs) in the reports to mask them and predict with our pre-trained CamemBERT model, which adjectives and adverbs could appropriately substitute the masked modifiers (except the original ones). We thereby generated a synthetic training dataset containing new modifiers which retained the same meaning and patterns. We trained the model again on the new training dataset that included both the original and the synthetic data. This technique was chosen because our model considers the context of the sentence that includes the masked word before predicting the appropriate word.

4   Results and Evaluation

We evaluated the quality of the summary using ROUGE [15]. Unigram and bi-gram overlap (R-1 and R-2) are reported as a means of evaluating informativeness and the longest common subsequence (R-L) as a means of evaluating fluency. We compared the performance of the different CamemBERT models, and we further included the Lead-3 [21] multilingual baseline that extracts the first 3 sentences from any given article and the PyTextRank [19], considered as multilingual baseline, which is an implementation that includes the TextRank [18], PositionRank [4] and Biased TextRank [9] algorithms, used for extractive summarisation by getting the top-ranked phrases from a text document.

Table 1. Fine-tuned CamemBERT model scores before and after data augmentation

Models	R_1	R_2	R_L
Lead-3 baseline	0.3342	0.1220	0.1735
PyTextRank baseline	0.3979	0.1269	**0.2616**
CamemBERT (Original)	0.4354	0.1056	0.2277
Fine-tuned CamemBERT (Without data augmentation)	**0.5301**	0.3238	**0.4203**
Fine-tuned CamemBERT (With data augmentation)	**0.6136**	0.4201	**0.5102**

Table 1 shows the significant advantage of training the model on our custom financial dataset, where the model learned a new domain-specific vocabulary, allowing us to generate more specific summaries. In addition, data augmentation on the Bank of France's textual monthly reports data contributed to an improvement of 6% to 11% in terms of the model's performance and the automatically-produced summaries.

5   Conclusion

The extraction of lexico-grammatical patterns from the attention mechanisms allowed our model trained on the bank reports to produce more accurate and domain-specific summaries. We have highlighted the impact of textual data augmentation on the model's performance and on the extracted summaries. The lack of data does not allow the model to cover all of the semantically relevant aspects of the domain-specific data. More original financial data could improve the model's performance.

References

1. Desmedt, L., Gautier, L., Llorca, M.: Les discours de la conjoncture économique. L'Harmattan, Paris (2021)

2. Devlin, J., Chang, M.W., Lee, K., Toutanova, K.: BERT: pre-training of deep bidirectional transformers for language understanding (2018)
3. Ettinger, A.: What BERT is not: lessons from a new suite of psycholinguistic diagnostics for language models. Trans. Assoc. Comput. Ling. **8**, 34–48 (2020)
4. Florescu, C., Caragea, C.: PositionRank: an unsupervised approach to keyphrase extraction from scholarly documents. In: Proceedings of the 55th Annual Meeting of the Association for Computational Linguistics, Vancouver, Canada (2017)
5. Gautier, L.: Figement et discours spécialisés. Frank und Timme, Berlin (1998)
6. Gautier, L.: Les discours de la bourse et de la finance. Frank und Timme, Berlin (2012)
7. Gledhill, C., Kübler, N.: What can linguistic approaches bring to English for specific purposes? ASp. la revue du GERAS **69**, 65–95 (2016)
8. Granger, S., Meunier, F.: Phraseology: An Interdisciplinary Perspective. John Benjamins Publishing, Amsterdam (2008)
9. Kazemi, A., Pérez-Rosas, V., Mihalcea, R.: Biased TextRank: unsupervised graph-based content extraction. In: Proceedings of the 28th International Conference on Computational Linguistics, Barcelona, Spain, pp. 1642–1652 (2020)
10. Kim, T., Choi, J., Edmiston, D., goo Lee, S.: Are pre-trained language models aware of phrases? Simple but strong baselines for grammar induction (2020)
11. Kudo, T., Richardson, J.: SentencePiece: a simple and language independent subword tokenizer and detokenizer for neural text processing (2018)
12. Laifa, A., Gautier, L., Cruz, C.: Extraire des patterns pour améliorer l'idiomaticité de résumés semiautomatiques en finances: le cas du lexique support. In: ToTh 2020 - Terminologie et Ontologie. Université Savoie Mont-Blanc, Presses Universitaires Savoie Mont-Blanc, Chambéry, France (2020)
13. Legallois, D., Charnois, T., Larjavaara, M.: The Grammar of Genres and Styles: From Discrete to Non-discrete Units. Walter de Gruyter GmbH & Co KG, Berlin (2018)
14. Legallois, D., Tutin, A.: Présentation: Vers une extension du domaine de la phraséologie. Langages (1), 3–25 (2013)
15. Lin, C.Y.: Rouge: a package for automatic evaluation of summaries. In: Text Summarization Branches Out, pp. 74–81 (2004)
16. Marecek, D., Rosa, R.: From balustrades to Pierre Vinken: looking for syntax in transformer self-attentions (2019)
17. Martin, L., et al.: Camembert: a tasty French language model (2019)
18. Mihalcea, R., Tarau, P.: TextRank: bringing order into text. In: Proceedings of the 2004 Conference on Empirical Methods in Natural Language Processing, pp. 404–411. Association for Computational Linguistics, Barcelona (2004)
19. Nathan, P.: PyTextRank, a Python implementation of TextRank for phrase extraction and summarization of text documents (2016)
20. Rocci, A., Palmieri, R., Gautier, L.: Introduction to thematic section on text and discourse analysis in financial communication. Stud. Commun. Sci. **15**(1), 2–4 (2015)
21. See, A., Liu, P.J., Manning, C.D.: Get to the point: summarization with pointer-generator networks. In: Proceedings of the 55th Annual Meeting of the Association for Computational Linguistics (Volume 1: Long Papers), pp. 1073–1083. Association for Computational Linguistics, Vancouver (2017)
22. Tenney, I., et al.: What do you learn from context? Probing for sentence structure in contextualized word representations (2019)
23. Vaswani, A., et al.: Attention is all you need. Adv. Neural. Inf. Process. Syst. **30**, 5998–6008 (2017)

24. Vig, J.: A multiscale visualization of attention in the transformer model. In: Proceedings of the 57th Annual Meeting of the Association for Computational Linguistics: System Demonstrations, pp. 37–42 (2019)
25. Vilares, D., Strzyz, M., Søgaard, A., Gómez-Rodríguez, C.: Parsing as pretraining. In: Proceedings of the AAAI Conference on Artificial Intelligence (2020)

Explainability in Irony Detection

Ege Berk Buyukbas[(✉)], Adnan Harun Dogan, Asli Umay Ozturk,
and Pinar Karagoz

Computer Engineering Department, Middle East Technical University,
Ankara, Turkey
{ege.buyukbas,adnan.dogan,ozturk.asli}@metu.edu.tr,
karagoz@ceng.metu.edu.tr

Abstract. Irony detection is a text analysis problem aiming to detect ironic content. The methods in the literature are mostly for English text. In this paper, we focus on irony detection in Turkish and we analyze the explainability of neural models using Shapley Additive Explanations (SHAP) and Local Interpretable Model-Agnostic Explanations (LIME). The analysis is conducted on a set of annotated sample sentences.

Keywords: Irony detection · Explainability · Sentiment analysis · Neural models · SHAP · LIME

1 Introduction

Irony detection on textual data is one of the most difficult subproblems of sentiment analysis. Irony is defined as *the expression of one's meaning by using language that normally signifies the opposite, typically for humorous or emphatic effect*[1] by the Oxford Dictionary. It is particularly a difficult problem since opposition of the meaning is mostly implicit, and automated emotion detection methods lack the understanding of *common sense* that we humans share. Irony detection on textual data can be considered as a classification task to determine whether the given text is either *ironic* or *non-ironic*. In this work, we focus on irony detection problem from *explainability* point of view, and we particularly explore the performance of neural models on Turkish text. In addition to BERT and LSTM, we adapt Text-to-Text Transfer Transformer (T5 [6]) for binary classification of irony.

Explainability is a popular topic on machine learning, aiming to provide insight for the predictions generated by models. Models generated by some of the algorithms, such as Decision Tree (DT), tend to explicitly provide such insight due to the nature of the algorithm. However, neural network based models and transformers generate *black-box models*. In this work, for explainability analysis, we use Shapley Additive Explanations (SHAP [4]) and Local Interpretable Model-Agnostic Explanations (LIME [7]) methods, both of which are model-agnostic, hence can be applied on any predictive model.

---

[1] https://www.lexico.com/en/definition/irony.

© Springer Nature Switzerland AG 2021
M. Golfarelli et al. (Eds.): DaWaK 2021, LNCS 12925, pp. 152–157, 2021.
https://doi.org/10.1007/978-3-030-86534-4_14

Explainability of text classification is an emerging topic with a limited number of studies using LIME and SHAP. In [1], layer-wise relevance propagation (LRP) has been used to understand relevant words in a text document. In [5], theoretical analysis of LIME for text classification problem is examined. According to their analysis, LIME can find meaningful explanations on simple models such as linear models and DTs, but the complex models are not fully analyzed.

As an important difference from such previous studies, we focus on irony detection problem, which carries further difficulty as it involves a non-standard use of natural language. As another difficulty, we conduct the analysis on a morphologically complex language, Turkish. For the analysis, we use a new irony data set in Turkish, which includes 600 sentences with balanced number of labels.

2 Methods

In this study, similar to the approaches used for English and Turkish in the literature [2,8,9], Bidirectional LSTM (Bi-LSTM) neural model is used. Additionally, the masked language model BERT [3][2] is used with the BERT Base Multilingual Cased pre-trained model and fine-tuned for the classification task [2]. Different from the previous studies, Text-to-Text Transformer (T5) [6][3] is also employed. T5 model is trained with an open-source pre-training TensorFlow dataset of about 7 TB, Colossal Clean Crawled Corpus[4] (C4).

The first explainability method we use is LIME [7], which is a local surrogate model. The basic idea behind the algorithm is perturbing a data instance, then establishing new predictions with those perturbed data from the black box model. By this way, LIME can explain decisions for instances locally. In this work, we use LIME Text Explainer package[5] for constructing and analyzing BERT and T5 models. For LIME explainer, the number of words in each sentence is used as the number of features. The number of perturbed samples is set as 5000, which is the default parameter. The second explainability analysis method is SHAP [4], which assigns importance values to features of the data. In our study, we use SHAP Kernel Explainer[6] with its default parameters for the analysis of the Bi-LSTM model. With SHAP, we can examine not only contribution of words, but also features of a data instance, such as *existance of exclamation mark (!)* or a *booster*.

3 Experiments and Results

For the analysis, we use IronyTR Extended Turkish Social Media Dataset[7], which consists of 600 sentences. Data set is balanced with 300 ironic and 300

---

[2] https://github.com/google-research/bert.
[3] https://ai.googleblog.com/2020/02/exploring-transfer-learning-with-t5.html.
[4] https://www.tensorflow.org/datasets/catalog/c4.
[5] https://github.com/marcotcr/lime.
[6] https://github.com/slundberg/shap.
[7] https://github.com/teghub/IronyTR.

Table 1. Comparison of methods on Turkish data set

Method	Accuracy	Precision	Recall	F1-score
LSTM	51.33%	55.09%	52.73%	53.88%
Bi-LSTM	50.16%	51.44%	62.07%	56.26%
BERT	**69.00%**	**71.34%**	**65.75%**	**68.43%**
T5	59.50%	59.21%	57.68%	56.85%

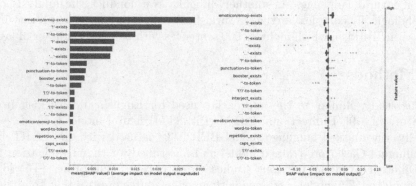

Fig. 1. Average and instance based feature impacts on Bi-LSTM Model (by SHAP method).

non-ironic instances. Sentences in the data set are collected from social media platforms either manually or by using the APIs of the platforms, and annotated by 7 Turkish native speakers to set the ground truth through majority voting.

For each model used for irony detection, we present accuracy, precision, recall, and F1-scores. All experiments are performed under 10-fold cross-validation. Table 1 gives the performance of LSTM, Bi-LSTM, BERT and T5, comparatively.

For explainability analysis, we applied SHAP method on Bi-LSTM model, LIME method on BERT and T5 models. The employed Bi-LSTM model uses feature-based inputs as well as text as sequence input. In order to analyze the effect of feature-based inputs, we preferred to apply SHAP for the analysis of this model.

We present a comparative explainability analysis of the models over two sample sentences, one with *irony* label, and the other with *non-irony* label. However, before this instance based comparison, we present the overall explainability analysis results by SHAP method on Bi-LSTM model.

General Analysis on Bi-LSTM by SHAP Method. In Fig. 1, the first graphics (on the left) show the average impacts of the features on the outcome as magnitudes, and the second graphics (on the right) shows the direct impacts. In the second graphics, the x-axis shows SHAP values of the features per instance. The orientations of the features' effects are displayed as well. In the figures, the most important feature on the model output is given as *emoticon/emoji-*

Table 2. Sentences, sentence numbers and translations.

Sent. #	Sentence
(a)	Hafta sonu meteor yağmurları gözlemlenebilecek.
transl.:	*Meteor showers can be observed at the weekend.*
(b)	Evde bayram temizliği yapacağız diye beni dışarı yolladılar, yolda fark ettim ki temizliğe benden başlamışlar çok üzgünüm şu an...
transl.:	*They sent me out to do holiday cleaning at home, I realized on the way that they started the cleaning with me, I'm very sad now...*
(c)	Yağmurlu havada su birikintilerinden hızlı geçmeye devam edin lütfen, ıslanmaya bayılıyoruz
transl.:	*Keep passing by through puddles fast in rainy weather please, we love to get wet.*

exists, which denotes if the sentence contains at least one emoji. According to the second graphics, if the sentence contains at least one emoji, the instance's prediction tends to be non-ironic, as the blue nodes denote that the sentence contains at least one emoji, whereas reddish nodes mean just the opposite. The second most important feature is *!- exists*, which denotes the sentence includes exclamation mark. According to the second graphics, if the sentence contains at least one exclamation mark, the instance's prediction tends to be ironic.

Analysis on Non-irony Sample. The non-irony sample we use is Sentence (a) in Table 2. According to Bi-LSTM model, all words in the sentence are effective for predicting the label as *non-irony*. The most effective words on the prediction are *yağmur (shower/rain)* and *meteor (meteor)* whose absolute SHAP values are maximum. Moreover, not only words but also the features '*!*'- *exists*, '*!*'- *to-token*, '*?*'- *to-token*, '*. . .*'- *to-token*, which are some of the most effective features given in Fig. 1, have similar effects on the model as their SHAP values are negative. The contribution of the words is higher than the features, as observed in the magnitude of the SHAP values. According to the results of the LIME method for BERT model, all the words in the sentence lead to *non-irony* label prediction as in Bi-LSTM. Moreover, *yağmur (shower/rain)* is the most effective word again, but the second most effective word is *gözlemlenebilecek (can be observed)* which is the least effective word in Bi-LSTM. The word *meteor (meteor)* is the third effective word. According to the result of the LIME method for T5 model analysis, *meteor (meteor)*, *yağmur (shower/rain)* and *gözlemlenebilecek (can be observed)* effect the label prediction as *non-irony*, as in BERT, but *hafta (at the week)* and *son (end)* affected the prediction oppositely. All three models predicted the sentence correctly as non-ironic.

Analysis on Irony Sample. For this analysis, we use Sentence (c) in Table 2. According to the results of SHAP method on Bi-LSTM, the following 7 words lead to prediction as *non-irony*: *yağmur (rain/shower)*, *hava (weather)*, *birikinti (puddle)*, *hızlı (fast)*, *geçmek (pass by)*, *devam (keep on)* and *ıslanmak (get wet)*.

Table 3. SHAP scores for Bi-LSTM, LIME scores for BERT and T5 models on the sample sentences.

Bi-LSTM	Snt #	# of pos.	Avg of pos.	Max pos.	# of neg.	Avg of neg.	Max neg.	Pred.
	(a)	0	–	–	5	−0,050	−0,081	TN
	(b)	9	0,046	0,244	0	–	–	TP
	(c)	3	0,029	0,033	7	−0,066	−0,098	FN
BERT	Snt #	# of pos.	Avg of pos.	Max pos.	# of neg.	Avg of neg.	Max neg.	Pred.
	(a)	0	–	–	5	−0,017	−0,028	TN
	(b)	6	0,094	0,225	10	−0,025	−0,048	TP
	(c)	3	0,032	0,043	8	−0,022	−0,054	FN
T5	Snt #	# of pos.	Avg of pos.	Max pos.	# of neg.	Avg of neg.	Max neg.	Pred.
	(a)	2	0,325	0,374	3	−0,308	−0,624	TN
	(b)	9	0,033	0,075	7	−0,008	−0,018	TP
	(c)	3	0,155	0,324	8	−0,158	−0,515	FN

On the other hand, the words *su (water)*, *etmek (to do/to make)*, and *lütfen (please)* have positive SHAP values, and drive the model to irony. Additionally, *'!'- exists*, *'!'- to-token*, *'?'- to-token*, *'. . .'- to-token* and *booster-exists* features in the sentence lead the prediction as non-irony, as their SHAP values are negative, whereas *emoticon/emoji-exists* has a positive SHAP value. The LIME method for BERT model indicates that every word in the sentence except for *su (water)* and *bayilmak (love to)* have the same effect on the model as in Bi-LSTM model. The word *su (water)* has an opposite effect on the BERT model, and the word *bayilmak (love to)* has an impact on the model to lead the prediction as irony, but overall prediction of BERT model is non-irony. According to result of LIME method for T5 model, given in Table 3, the only words that lead to irony label prediction are *birikinti (puddle)*, *lütfen (please)* and *bayilmak (love to)*. However, the sentence is misclassified again as the other words contribute to non-irony for label prediction. Therefore, although the three models captured some words that lead to irony, they all mispredicted the sentence as non-irony.

Analysis on Sample Sentences. In order to give a more general view, we sampled 3 sentences as given in Table 2, where the first sentence is non-ironic, and the rest are ironic. The summary of SHAP results statistics and class label predictions as given in Table 3 for Bi-LSTM model. Similarly, summary of LIME weights and irony label predictions for BERT and T5 are presented in Table 3 and Table 3, respectively. For each prediction model, we observe slightly different effects of words. Considering the sentence (a), which is correctly labeled as *non-irony* by all three models, it is seen that in Bi-LSTM and BERT, all the words contribute to the correct label prediction, whereas in T5, although 2 of the words contribute to *irony* class label, 3 of the words determine the class label, especially one word with the maximum magnitude of −0,624. As another example, for the sentence (b), which is correctly labeled as *irony* by all models, in Bi-LSTM, again, all the words contribute to the correct label prediction. On the other hand, in BERT, the majority of the words lead to incorrect labels with small weights, whereas there is a word with the maximum weight of 0,225 to determine the class label as irony. As for T5, the behavior is similar but not as strong as in BERT. It is seen that 7 out of 16 words have an effect towards non-irony label, and the effect of the other 9 words lead to the correct prediction with a maximum weight of 0,075. Overall, we can conclude that the effect towards the label prediction is distributed over a set of words in Bi-LSTM, yet we see few words providing a stronger effect for prediction in the transformer models.

4   Conclusion

In this paper, we investigate the explainability of LSTM, Bi-LSTM, BERT, and T5 based irony detection models on Turkish informal texts using SHAP and LIME methods. In terms of explainability, our analysis shows that, as expected, usage of punctuations such as *"!"*, *"(!)"* and *"..."* is a sign of irony in the detection models. The contribution of the words to the label prediction slightly differs for the models. In Bi-LSTM, many of the words in a sentence contribute to the prediction with comparatively smaller weights. On the other hand, for BERT and T5, fewer number of strong words determine the class label. As future work, using a multi-lingual model for T5 may be considered for irony detection performance and explainability analysis.

References

1. Arras, L., Horn, F., Montavon, G., Müller, K.R., Samek, W.: "What is relevant in a text document?": an interpretable machine learning approach. PLoS ONE **12**(8) (2017)
2. Cemek, Y., Cidecio, C., Ozturk, A.U., Cekinel, R.F., Karagoz, P.: Turkce resmi olmayan metinlerde ironi tespiti icin sinirsel yontemlerin incelenmesi (investigating the neural models for irony detection on turkish informal texts). IEEE Sinyal Isleme ve Iletisim Uygulamalari Kurultayi (SIU2020) (2020)
3. Devlin, J., Chang, M.W., Lee, K., Toutanova, K.: Bert: pre-training of deep bidirectional transformers for language understanding. arXiv preprint arXiv:1810.04805 (2018)
4. Lundberg, S.M., Lee, S.I.: A unified approach to interpreting model predictions. In: Guyon, I., Luxburg, U.V., Bengio, S., Wallach, H., Fergus, R., Vishwanathan, S., Garnett, R. (eds.) Advances in Neural Information Processing Systems, vol. 30, pp. 4765–4774. Curran Associates, Inc. (2017)
5. Mardaoui, D., Garreau, D.: An analysis of lime for text data. In: International Conference on Artificial Intelligence and Statistics, pp. 3493–3501. PMLR (2021)
6. Raffel, C., et al.: Exploring the limits of transfer learning with a unified text-to-text transformer. arXiv preprint arXiv:1910.10683 (2019)
7. Ribeiro, M.T., Singh, S., Guestrin, C.: "Why should i trust you?" explaining the predictions of any classifier. In: Proceedings of the 22nd ACM SIGKDD International Conference on Knowledge Discovery and Data Mining, pp. 1135–1144 (2016)
8. Wu, C., Wu, F., Wu, S., Liu, J., Yuan, Z., Huang, Y.: THU-NGN at SemEval-2018 task 3: tweet irony detection with densely connected LSTM and multi-task learning. Proceedings of the 12th International Workshop on Semantic Evaluation (SemEval-2018), pp. 51–56 (2018)
9. Zhang, S., Zhang, X., Chan, J., Rosso, P.: Irony detection via sentiment-based transfer learning. Inf. Process. Manage. **56**(5), 1633–1644 (2019)

Efficient Graph Analytics in Python
for Large-Scale Data Science

Xiantian Zhou[✉] and Carlos Ordonez

University of Houston, Houston, TX 77204, USA

Abstract. Graph analytics is important in data science research, where Python is nowadays the most popular language among data analysts. It facilitates many packages for graph analytics. However, those packages are either too specific or cannot work on graphs that cannot fit into the main memory. Moreover, it is hard to handle new graph algorithms or even customize existing ones according to the analyst's need. In this paper, we develop a general graph C++ function based on a semiring algorithm including two math operators. The function can help solve many graph problems. It also works for graphs that cannot fit in the main memory. Our function is developed in C++, but it can be easily called in Python. Experimental comparison with state-of-art Python packages show that our C++ function has comparative performance for both small and large graphs.

Keywords: Graph analysis · Data science · Python · Template algorithm

1 Introduction

Graph analytics remains one of the most computationally intensive tasks in data science research mainly due to large graph sizes and the structure of the graphs. On the other hand, Python is the most popular system to perform data analysis because of its ample library of models, powerful data transformation operators, interpreted and interactive language. However, Python is slow to analyze large data sets, especially when data cannot fit in RAM. Many research progress on efficient analytic algorithms working on graph analytics engines, which frequently needs data exporting and importing [3,4]. But exporting data sets from or to a graph engine is slow and redundant. Even though many libraries in Python enable graph analytics, these libraries require time to learn, or users need re-programming existing functions, which limits their impact [1,5,6]. Moreover, graph functions provided by Python packages are too specific. If a user needs a new graph algorithm or customizes existing ones, none of the Python packages can be helpful. Graph analytics highly depends on the APIs that packages provide. However, graph analytics is a rapidly developing research field, in which there are many new graph metrics and algorithms appearing.

© Springer Nature Switzerland AG 2021
M. Golfarelli et al. (Eds.): DaWaK 2021, LNCS 12925, pp. 158–164, 2021.
https://doi.org/10.1007/978-3-030-86534-4_15

In this paper, we prove that it is possible to identify a general graph C++ function that can solve lots of graph algorithms. we develop the C++ function which is the most computationally intensive part in many graph analytics with C++. The function is light-weighted and can be easily called in Python. Also, it can efficiently work for both small and large graphs which cannot fit in the main memory. Moreover, it is convenient to customize a graph algorithm or program a new one with our function.

2   Related Work

Graph analytics in Python is becoming more important in data science research. However, it is challenging because of large graph size and complex patterns embedded in the graph. There are many graph packages developed recently. Sciki-network is a package based on SciPy, and it provides state-of-the-art algorithms for ranking, clustering, classifying, embedding and visualizing the nodes of a graph [1]. Python-iGraph is a library written in C++, which provides an interface to many graph algorithms [5]. The graph-tool is also an efficient package that is developed in C++ and it can works in multi-threads. GraphBLAS is another popular packages, but it needs much time to learn [2,6]. Most of those libraries cannot work for graphs that can not fit in the main memory. Also, they provide too specific interfaces that we can not modify or customize graph algorithms.

3   Definitions

Let $G = (V, E)$ be a directed graph with $n = |V|$ vertices and $m = |E|$ edges, where V is a set of vertices and E is a set of edges. An edge in E links two vertices in V, has a direction and a weight. The adjacency matrix of G is a $n \times n$ matrix such that the entry i, j holds 1 when exists an edge from vertex i to vertex j. In sparse graph processing, it is preferable to store the adjacency matrix of G in a sparse form, which saves spaces and CPU resources. There are many different sparse formats, such as compressed sparse row, compressed sparse column, and coordinate list, and other data structures. In our work, the input and output are represented as a set of tuples (i, j, v) where i and j represent the source and the destination vertex, and v represents the value of edge (i, j). The set of tuples can be sorted by j or i. We denote E_i is a set of tuples sorted by i and E_j is sorted by j. Since entries where $v = 0$ are not stored, the space complexity is $m = |E|$. In sparse matrices, we assume $m = O(n)$.

4   A Graph Analytics Function Based on a Semiring

We start by explaining classical graph problems, from which we derived the general graph analytics function which is an algorithm of a semiring including two math operators. Then, we introduce the architecture of the function. Finally, we use it to implement different graph algorithms and show the function can solve many graph problems.

4.1   A General Graph Analytics Function

Graph metrics are about either local information or global information of a graph. Some graph metrics such as in-degree and out-degree are about local information. Many other graph metrics such as PageRank, centrality, reachability, and minimum spanning tree are about global information. Those graph metrics require a part or full traversal of the graph. There are different ways to obtain local information of a graph, and adjacent matrix-matrix or matrix-vector operations [8] is one of the most commonly used solutions. For example, the matrix product $E \cdot S$, where S is an n-dimensional vector, helps finding the vertices with highest connectivity. $E \cdot E$ is used to find paths whose lengths are two.

To obtain the global information, we can repeat the matrix-matrix or matrix-vector multiplication process. For example, the matrix product $E \cdot E \cdot E$ can be used to get all triangles in G, which has been identified as an important primitive operation. The iteration of $E \cdot E$, multiplying E k times (k up to $(n-1)$) gets all paths with length k, from which we can filter the shortest/longest ones by different aggregation operators and count them. Also, it gets all the intermediate vertices on those paths in order, which is essential for path analysis, such as betweenness centrality. Finally, $E \cdot E \cdot ...E$ ($n-1$ times) until a partial product vanishes is a demanding computation returning $G+$, the transitive closure (reachability) of G which gives a comprehensive picture about G connectivity [7]. Besides connectivity and paths, other graph metrics such as PageRank, closeness can also be computed as powers of a modified transition matrix. Note that although we express those graph algorithms as matrix multiplication, they are different semirings according to different graph algorithms. Semirings are algebraic structures defined as a tuple $R(R, \oplus, \otimes, 0, 1)$ consisting of a set R, an additive operator $\oplus$ with identity element 0, a product operator $\otimes$ with identity element 1. The regular matrix multiplication is defined under $(R, +, \times, 0, 1)$. A general definition of matrix multiplication expands it to any semiring. For example, (min, add) is used to solve shortest paths problem where min is the additive operator, and add is the product operator. The boolean semiring, with $\vee$ (logical OR) as the additive operator and $\wedge$ (logical AND) as the product operator, is frequently used in linear algebra to represent some graph algorithms. We use $\oplus, \otimes$ to denote the additive and product operator in a semiring.

We can see that all algorithms mentioned above can be expressed as a matrix-vector or matrix-matrix operation or an iteration of such operations under different semirings. The general algorithm is shown in Algorithm 1. We show it based on a dense matrix multiplication to represent our general graph algorithm. The number of loops can vary according to different graph algorithms. Based on this mathematical foundation, we develop a general graph C++ function $s(E, \oplus, \otimes, K)$ in Python which takes E as input and performs computation according to two operators $(\oplus, \otimes)$ of a semiring. For some graph algorithms, we only want to traverse parts instead of the entire graph, especially when the graph is large. So we use k to specify the maximum number of iterations. Intermediate vertices are needed for graph metrics such as betweenness centrality,

Input: $E, V, \oplus, \otimes$
Output: R
1 $n \leftarrow |V|$;
2 for $i \leftarrow 1$ to n do
3 | for $j \leftarrow 1$ to n do
4 | | for $k \leftarrow 1$ to n do
5 | | | $R[i,j] = R[i,j] \oplus (E[i,k] \otimes E[k,j])$
6 | | end
7 | end
8 end

Algorithm 1: General graph algorithm

k-step betweenness centrality. In contrast, they are not required for other graph metrics such as reachability and connectivity. So we add an optional boolean parameter h which indicates whether intermediate vertices on paths need to be recorded or not. The default value of h is false.

Input: $s(E, \oplus, \otimes, K)$
Output: R
1 $k \leftarrow 0, i \leftarrow 1, j \leftarrow 1, E_i \leftarrow E$ sorted by i, $E_j \leftarrow E$ sorted by j
   for $k < K$ do
2 | while *not end of E_i and not end of E_j* do
3 | | read a block R_b if $k \neq 0$;
4 | | read a column block B_j from E_j with j maximum index is j_B;
5 | | read a row block B_i from E_i with i maximum index is i_B; **while** $i_t < i_B, j_t < j_B$ do
6 | | | **foreach** *pair $i_t = j_t$* do
7 | | | | $R_b[i,j] = R_b[i,j] \oplus (B_i[i,j_t] \otimes B_j[i_t,j])$;
8 | | | end
9 | | end
10 | | periodically write the $R_b[i,j]$ into disk;
11 | end
12 | Merge all $R_b[i,j]$ into R;
13 | $k \leftarrow k+1$
14 end

Algorithm 2: The database algorithm of our graph C++ function

Now we show how to do graph analytics with our function. By specifying the function as $s(E, min, add, n)$, we can calculate shortest paths, where k equals n because of traversing the full graph. Similarly, the reachability of all vertices can be calculated with $s(E, \vee, \wedge, n)$. Using $s(E, add, mul, n)$, we can obtain the number of paths between each pair of vertices. If we want to calculate K-step betweenness centrality which is not included in many Python packages, we can use $s(E, min, add, K, h = true)$ get all shortest paths in K steps with interme- diate vertices along them. Even for many other graph metrics other than paths and connectivity, such as PageRank and closeness, the C++ function can also be used to calculate the most computation intensive part. The function triggers computation in a C++ environment. The results are written to disk. The architecture of the function is discussed in the following section.

4.2   System Architecture for Function Processing

Our goal is to provide a general function running on a simple architecture to do graph analytics in a data science language. The function is developed in C++ environment to achieve good performance. And we divide the input files into blocks and process block by block when the computation involves a large graph, so the function works for graphs that cannot fit in the main memory.

Input and Output: The graph is stored in (i, j, v) format, and we assume that the input files are text files that are stored in disk. It can be in any text file format like .csv, .txt, and so on. Before processing, we first sort the input by i, then by j. The output file is also in (i, j, v) format and sorted by i.

Processing: When our function is called in Python, the computation process in C++ will be triggered. It reads files from disk, performs computations according to the input arguments specified in the function input, and writes the final results into a text file format. If the graph is too big to fit in the main memory, the C++ function will read and process input files block by block. The algorithm is shown in Algorithm 2. Remember that E_i is sorted by i and E_j is sorted by j.

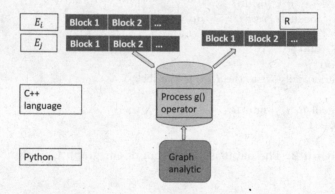

Fig. 1. An overview of our system

5   Experimental Evaluation

NumPy and SciPy are fast Python packages for processing matrices. SciPy has a high performance and low memory use achieved through a mix of fast sparse matrix-vector. Those two packages have the basic matrix-matrix multiplication operator. Thus, we choose them to compare with our C++ function. The computer used for the experiments has an Intel Pentium(R) CPU running at 1.6 GHz, 8 GB of RAM, 1TB disk, 224 kb L1 cache, 2 MB L2 cache and running Linux Ubuntu 14.04. We use both synthetic and real graph data sets for experimental evaluation. The real graphs are from the Stanford SNAP repository. The comparison results are shown in Table 1. If the running time is more than one hour, we put a "Stop" sign in the table. From Table 1, we can see that our function is faster than NumPy, but it is slower than SciPy for small graphs. However, SciPy crashes for large graphs because it assumes data can fit in the main memory. Moreover, our function can perform matrix operations under different semirings while most other packages can only perform matrix multiplication.

Table 1. Summary of results (time in seconds).

Dataset	Type	n	m	NumPy	SciPy	Our C++ function
synthgraph1	Synthetic	1.0K	100.0K	12.0	0.2	1.0
synthgraph2	Synthetic	5.0K	2.5M	Stop	14.0	162.0
wiki-vote	Real	8.0K	103.6K	Crashed	0.1	1.6
webgoogle	Real	875.0K	5.1M	Crashed	Crashed	Stop

6   Conclusions

In this paper, we developed a powerful, yet easy to use function based on a semiring algorithm running on an efficient system architecture to process big graphs in a data science language. The function can be easily called in Python, and it can be used for programming, modifying and customizing most of existing and new graph algorithms. Moreover, it works for large graphs that cannot fit in the main memory. We compared the performance of our function with state-of-art graph analytic packages. The results show that our C++ function has comparative performance for small graphs and large graphs that cannot fit in the main memory.

For future work, we will study parallel processing, develop more efficient algorithms when results fit in main memory (e.g. PageRank, Connected Components). Finally, we plan to characterize which graph algorithms pattern fits our semiring algorithm.

References

1. Bonald, T., de Lara, N., Lutz, Q., Charpentier, B.: Scikit-network: graph analysis in Python. J. Mach. Learn. Res. **21**, 185:1–185:6 (2020)
2. Chamberlin, J., Zalewski, M., McMillan, S., Lumsdaine, A.: PyGB: GraphBLAS DSL in Python with dynamic compilation into efficient C++. In: IPDPS Workshops 2018, pp. 310–319 (2018)
3. Ghrab, A., Romero, O., Jouili, S., Skhiri, S.: Graph BI & analytics: current state and future challenges. In: Ordonez, C., Bellatreche, L. (eds.) DaWaK 2018. LNCS, vol. 11031, pp. 3–18. Springer, Cham (2018). https://doi.org/10.1007/978-3-319-98539-8_1
4. Ho, L., Li, T., Wu, J., Liu, P.: Kylin: an efficient and scalable graph data processing system. In: 2013 IEEE Big Data, pp. 193–198 (2013)
5. Ju, W., Li, J., Yu, W., Zhang, R.: iGraph: an incremental data processing system for dynamic graph. Front. Comput. Sci. **10**(3), 462–476 (2016). https://doi.org/10.1007/s11704-016-5485-7
6. Kumar, M., Moreira, J.E., Pattnaik, P.: GraphBLAS: handling performance concerns in large graph analytics. In: 2015, ACM CF, pp. 260–267. ACM (2018)
7. Ordonez, C., Cabrera, W., Gurram, A.: Comparing columnar, row and array DBMSs to process recursive queries on graphs. Inf. Syst. **63**, 66–79 (2016)
8. Zhou, X., Ordonez, C.: Computing complex graph properties with SQL queries. In: 2019 IEEE Big Data, pp. 4808–4816. IEEE (2019)

Machine Learning and Deep Learning

A New Accurate Clustering Approach for Detecting Different Densities in High Dimensional Data

Nabil El Malki$^{(\boxtimes)}$, Robin Cugny, Olivier Teste, and Franck Ravat

IRIT, Toulouse, France
nabil.el-malki@irit.fr

Abstract. Clustering is a data analysis method for extracting knowledge by discovering groups of data called clusters. Density-based clustering methods have proven to be effective for arbitrary-shaped clusters, but they have difficulties to find low-density clusters, near clusters with similar densities, and clusters in high-dimensional data. Our proposal consists in a new clustering algorithm based on spatial density and probabilistic approach. Sub-clusters are constituted using spatial density represented as probability density function ($p.d.f$) of pairwise distances between points. To agglomerate similar sub-clusters we combine spatial and probabilistic distances. We show that our approach outperforms main state-of-the-art density-based clustering methods on a wide variety of datasets.

Keywords: Density-based clustering · Probability density functions · Wasserstein distance

1 Introduction

Clustering methods are popular techniques widely used in machine learning to extract knowledge from a variety of datasets in numerous applications [1]. Clustering methods group similar data into a subset known as cluster. The clustering [1] consists in partitioning a dataset annotated $X = \{x_1, ..., x_n\}$ with $n = |X|$ into c clusters $C_1, ..., C_c$, so that $X = \cup_{i=1}^{c} C_i$. We only consider hard clustering according to which $\forall i \in [1..c], \forall j \neq i \in [1..c], C_i \cap C_j = \varnothing$.

In this paper, we focus on density-based clustering methods [2] that are able to identify clusters of arbitrary shapes. Another interesting property of these approaches is that they do not require the user to specify the number of clusters c. Density-based clustering is based on the exploration of high concentrations (density) of points in the dataset [3]. In density-based clustering, a cluster in a data space is a contiguous region of high point density separated from other such clusters by contiguous regions of low point density [2]. Nevertheless, density-based clustering has difficulties to detect clusters having low densities regarding high-density clusters. Low-density points are either considered as outliers or

© Springer Nature Switzerland AG 2021
M. Golfarelli et al. (Eds.): DaWaK 2021, LNCS 12925, pp. 167–179, 2021.
https://doi.org/10.1007/978-3-030-86534-4_16

included in another cluster. Moreover, near clusters of similar densities are often grouped in one cluster. Finally, density-based clustering does not manage well in high-dimensional data [4]. This paper tackles these challenging issues by defining a new density-based clustering approach.

Among the most known density-based algorithms are DBSCAN [5], OPTICS [6], HDBSCAN [7], DBCLASD [8], and DENCLUE [9]. Historically first, [5] introduces density as a minimum number of points within a given radius to discover clusters, but it poorly manage clusters having different densities. [6] addresses these varying density clusters by ordering points according to a density measure. [7] improved the approach by introducing a new density measure and an optimization function aiming at finding the best clustering solution. Although these approaches solve part of the problem of varying density clusters, they suffer from unevenly distributed density and high-dimensional datasets. They still mismanage low-density clusters by tending to consider them as outliers or to merge them into a higher-density cluster. [8] introduces a probabilistic approach of the density. It assumes that clusters follow a uniform probability law allowing it to be parameter-free. However, it has difficulties to detect non-uniform clusters because of this strong assumption. As [8] is a grid-based approach, its density calculations are less precise when the dimension increases [3]. Finally, [9] detects clusters using the probability density function $(p.d.f)$ of points in data space. [9] generalizes DBSCAN [5] and K-means [9] approaches. Therefore it inherits of their difficulties to detect different densities and low-density clusters. The common problems of the existing density-based clustering approaches are related to the difficulty of handling low-density clusters, near clusters of similar densities, and high-dimensional data. The other limit concerns their inefficiency to support nested clusters of different shapes and uneven distribution of densities.

We propose a new clustering approach that overcomes these limitations. The contributions of this article are as follows: (1) We define a new way to divide a dataset into homogeneous groups of points in terms of density. (2) We propose a new agglomerative approach for clustering based on Wasserstein metric and pairwise distance $p.d.f$ of sub-clusters. (3) We propose DECWA (DEnsity-based Clustering using WAsserstein distance), a hybrid solution combining density and probabilistic approaches. (4) We conducted experiments on a wide variety of datasets, showing that DECWA outperforms well known density-based algorithms in clustering quality by an average of 23%. Also, it works efficiently in high-dimensional data comparing to the others.

2   DEnsity-Based Clustering Using WAsserstein Distance

Our approach considers a cluster as a contiguous region of points where density follows its own law of probability. We represent a cluster C_i by a set of pairwise distances between the points contained in the cluster C_i. This set is annotated D_i and follows any law of probability. D_i is computed via any distance $d : X \times X \to \mathbb{R}^+$ that has to be at least symmetric and reflexive.

Density-based clustering methods consider the dataset X defined in coordinate space as a sample of a unknown probability density function. This function is a mixture of density functions, each corresponding to a group of points. The purpose of these methods is to estimate this function to extract clusters. DECWA also integrates this notion but applied to the space of pairwise distances instead of the coordinate space that is usually used in this kind of method [10]. We consider the distance space (associated with its coordinate space X) described by the probability density function of distances. Reasoning on distances makes it possible to detect different densities without suffering from the high dimensionality [3]. Indeed, a subset of distances represents several separated areas of similar densities in X. In Fig. 1, both representations are shown. Fig. 1(a) is the coordinate space representation of Compound dataset as well as Fig. 1(b) is its corresponding representation using the frequency of pairwise distances where each color represents a level of density. In Fig. 1(a) the same density level is shared by several separated groups of points in the coordinate space.

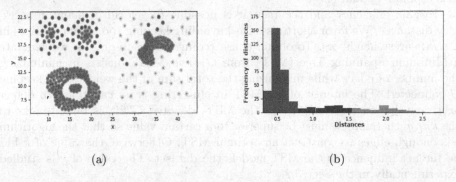

(a)   (b)

Fig. 1. a) Compound, a dataset with clusters of different densities, b) Histogram representation of the distances in the MST for Compound dataset.

The proposed solution named DECWA consists of four consecutive steps:

1. The first step transforms the dataset X into a k-nearest neighbor graph ($knng$) representation where nodes are points and edges are pairwise distances. The graph is then reduced to its Minimum Spanning Tree (MST) in order to keep significant distances.
2. The second step consists in calculating the $p.d.f$ from the significant distances of the MST. This $p.d.f$ is used to determine the extraction thresholds used to separate the different groups of spatial densities.
3. The third step consists of extracting sub-graphs from the MST according to extraction thresholds and identifying corresponding sub-clusters homogeneous in terms of spatial density.
4. The fourth step is to agglomerate sub-clusters according to spatial and probabilistic distances. Here, Wasserstein distance measures the similarity between probability distributions to regroup similar clusters.

2.1 Graph-Oriented Modeling of the Dataset

To estimate the distance $p.d.f$, we first calculate distances between the points of the dataset. Our computation strategy of pairwise distances is based on the k-nearest neighbor method [3]. The k-nearest neighbor graph highlights the local relationships between each point and its surrounding points in the data space. By extension, these relations make it possible to better capture the different spatial densities with local precision.

Therefore, the first step is the construction of the k-nearest neighbor graph of the dataset. From the dataset X, an undirected weighted graph annotated $G = (V, E)$ is constructed, where $V = \{v_1, v_2, ...\}$ is the set of vertices representing data points, and $E = \{e_q = (v_x, v_y, w_q) | v_x \in V, v_y \in V, w_q \in \mathbb{R}^+\}$ is the set of edges between data points. For each point, we determine k edges to k-nearest neighbor points. The weight w_q of each edge e_q is the distance between the two linked points v_x and v_y.

To gain efficiency and accuracy, it is possible to get rid of many unnecessary distances. We favor short distances, avoiding keeping too large distances in the k-nearest neighbors. To obtain dense (connected) sub-graphs, we generate a Minimum Spanning Tree (MST) from G. The MST consists in minimizing the number of edges while minimizing the total sum of the weights and keeping G connected. The interest of the MST in clustering is its capability to detect arbitrary-shaped clusters [11]. As the MST, denoted G^{min}, is built thanks to the $knng$, therefore, k must be superior to a certain value so that the algorithm has enough edges to construct an optimal MST. Otherwise, the value of k has no further influence as the MST models the dataset. The value of k is studied experimentally in the Sect. 3.2.

2.2 Probability Density Estimation

The overall objective of the $p.d.f$ estimation is to extract new information to delimit sub-clusters. Density-based methods search for the densest areas from a $p.d.f$ defined as $f : X \rightarrow \mathbb{R}^+$. They correspond in f to the areas around the peaks (included) called level sets [10]. Let $l > 0$ be a density threshold, a level set is a set of points defined as $\{x_i | f(x_i) > l\}$. This threshold l is on both sides of the bell (level set) centered around the peak. The sets of points $\{x_i | f(x) \leq l\}$ are either considered as outliers or assigned to denser clusters. This approach is not suitable for capturing clusters of different densities. Another solution [10], based on the hierarchical approach, cuts f at different levels l to better capture clusters of different densities. In our case, f is defined as $f : D' \rightarrow \mathbb{R}^+$ with D' the set of distances. We consider in our case that a level set has two thresholds that are not necessarily equal. Areas of f below these thresholds are not considered as outliers. They are the less frequent distances that correspond to phenomenons with fewer points such as overlapping areas or small clusters.

For estimating the $p.d.f$, we use the kernel density estimation (KDE) [12]. It makes no a priori hypothesis on the probability law of distances. The density function is defined as the sum of the kernel functions K on every distance. DECWA uses the Gaussian kernel. The KDE equation is below:

$$\widehat{f_h}(a) = \frac{1}{(n-1)h} \sum_{i=1}^{n-1} K\left(\frac{a - a_i}{h}\right)$$

With a and a_i that correspond to the distances in G^{min}; $n = |X|$ is the number of nodes, and $n - 1$ is the number of distances in G^{min}. The smoothing factor $h \in [0; +\infty]$, called the bandwidth, acts as a precision parameter that influences the number of peaks detected in the $p.d.f$ and therefore, the number of different densities.

2.3 Graph Division

To separate different densities, the next step is to find where to cut the $p.d.f$. The significantly different densities are detected with the peaks of the distance $p.d.f$ curve. We consider an extraction threshold as the mid-distance between each maximum and its consecutive minimum, and conversely between each minimum and its consecutive maximum. Mid-distances capture regions that are difficult to detect (low-density regions, dense regions containing few points, and overlapping regions that have the same densities). These regions are often very difficult to capture properly, whilst our mid-distance approach makes it possible to detect these cases. Once the mid-distances (ordered list called $\mho$) are identified on the $p.d.f.$ curve, we apply a process to treat successively the list of mid-distances in descending order. We generate sub-clusters, from the nodes of G^{min}, and sets of distances of sub-clusters, from the edges of G^{min}.

From G^{min}, we extract connected sub-graphs $\wp_t$ where all the nodes that are exclusively linked by edges greater or equal than the current mid-distance l_t. A node linked to an edge greater or equal to the l_t and another edge less than the l_t is not included. A sub-cluster C_i is composed of points that belong to one connected sub-graph. For each sub-cluster, we produce its associated set of distances D_i from edges between nodes of the sub-graph. A residual graph $\aleph_t$ is made up of edges whose distances are less than l_t, and all the nodes linked by these edges. This residual graph is used in the successive iterations. The residual graph can consist of several unconnected sub-graphs. At the last iteration (i.e. at the level of l_1), the residual graph $\aleph_1$ is also used to generate sub-clusters and associated sets of distances. After this step, some sub-graphs are close to each other while having a similar $p.d.f.$ According to our cluster characteristics, they should be merged. Figure 2 shows a comparison of the division result and the ground truth clusters for jain dataset; (a) shows the result of the division and (b) is the ground truth. The next step consists in identifying relevant sub-clusters to be agglomerated.

The connected sub-graphs of G^{min} have edge weights that are homogeneous overall. Indeed, the weight values of a connected sub-graph, annotated sg_i are framed between consecutive thresholds $l_t \in \mho, l_{t+1} \in \mho$, which guarantees its homogeneity in terms of spatial density.

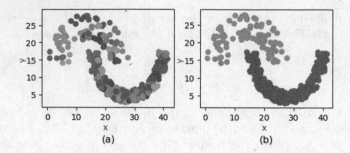

Fig. 2. Comparison of the division result with actual clusters for jain dataset

---

Algorithm 1. Graph division process

Input: $G^{min} = (V^{min}, E^{min}), \mho = \{..., l_t, l_{t+1}, ...\}$

Output: $\wp_1$

1: $\wp_{|\mho|} \leftarrow \emptyset$
2: $\aleph_{|\mho|} \leftarrow (V^{min}, E^{min})$
3: **for** $t \leftarrow |\mho| - 1$ **to** 1 **do**
4:     $E_t \leftarrow \{e_q \in E_{t+1} | w_q < l_t\}$
5:     $V_t \leftarrow \{v_q \in V_{t+1} | \exists (v_q, v_y, w_q) \in E_t \vee (v_x, v_q, w_q) \in E_t\}$
6:     $\aleph_t \leftarrow (V_t, E_t)$
7:     $\wp_t \leftarrow \wp_{t+1} \cup \{(V_{t+1} \setminus V_t, E_{t+1} \setminus E_t)\}$
8: **end for**
9: $\wp_1 \leftarrow \wp_1 \cup \aleph_1$

To show with more details the above statement, let us consider the list of mid-distances $\mho$, and a connected sub-graph sg_i with the weighted edges E_i. Suppose that $\mho$ is iterated consecutively in descending order as illustrated in Algorithm 1. In a given iteration t, the current threshold is l_t. One of the two following situations occurs: either $\forall e_q \in E_i, w_q < l_t$, in which case no action is taken on sg_i; or else if $\exists e_q \in E_i, w_q \geq l_t$ then from sg_i is extracted a set of connected sub-graphs called $\aleph_t$ with $w_q < l_t$ and another set of connected sub-graphs $\wp_t$ with $w_q \geq l_t$. At this point, for all edges into $\aleph_t$, $w_q \in]0; l_t[$. At the next iteration, $\aleph_t$ undergoes the same processing for the threshold l_{t-1}. If $\exists w_q \geq l_{t-1}$, the set of sub-graphs $\wp_{t-1}$ is extracted from $\aleph_t$. In this case, for all edges into $\wp_{t-1}$, $l_{t-1} \leq w_q < l_t$.

2.4   Agglomeration of Sub-clusters

This step aims at generating clusters from sub-cluster agglomeration. The main difficulty is to determine which sub-clusters should be merged. When two sub-clusters are close, it is difficult to decide whether or not to merge them. To arbitrate this decision, we use their density, represented as the distance $p.d.f$. Only sub-clusters with similar distance $p.d.f$ will be merged. The agglomeration process consists in merging every sub-clusters C_i and C_j that satisfy two conditions.

The first condition is that C_i and C_j must be spatially close enough. To ensure this, $d(C_i, C_j) \leq \lambda$ must be respected, with $\lambda \in \mathbb{R}^+$. It is necessary to determine the distance between sub-clusters ($d(C_i, C_j)$) to verify this condition. However, calculating every pairwise distance between two sub-clusters is a time-consuming operation. We consider that the value of the edge linking sub-graphs corresponding to C_i and C_j as the distance between C_i and C_j. Because of the MST structure, this distance is nearly the minimum distance between the points of C_i and C_j.

The second condition relates to the similarity of the distance distributions D_i and D_j. The purpose of this condition is to ensure $ws(D_i, D_j) \leq \mu$, with $\mu \in \mathbb{R}^+$, and ws the Wasserstein distance. We have opted for the Wasserstein distance as a measure of similarity between probability distributions. Two advantages led us to its choice [13]; (1) it takes into account the geometric shape of the $p.d.f.$ curves, and (2) if the two distributions are very close, then the small difference will be captured by Wasserstein, allowing a fine precision.

We introduce an iterative process for merging sub-clusters. First of all, it identifies the edge set $E_g \subset E^{min}$ containing only nodes whose respective sub-graphs are different. Then it runs an iterative sub-process consisting in iterating on the edges of E_g. For each edge $e_q \in E_g$ we verify that $d(C_i, C_j) \leq \lambda$ and $ws(D_i, D_j) \leq \mu$. In this case, C_i and C_j are merged. This is operated by the union of D_i, D_j and w_q and considering the points of C_i and C_j as belonging to the same sub-graph. Note that if $min(|D_i|, |D_j|) \leq 1$ then only the first condition applies as ws can not be calculated with this cardinality. The iterative sub-process is repeated until there are no longer merges. At last iteration the points of the clusters whose cardinality is 1 are considered as outliers.

2.5   Making Hyperparameters Practical

In the merging phase, λ and μ are metric thresholds for d and ws. Therefore they are defined in $]0, +\infty[$. It is difficult for users to operate with this unbounded interval. In this sub-section, the objective is to map this interval to a bounded interval defined between 0 and 1. We argue that a bounded interval is more meaningful for users. We consider $m \in \mathbb{R}^+, M \in \mathbb{R}^+$ respectively the lower bound and the upper bound of D' the set of distances and $\gamma, \tau \in]0, 1]$, so λ, μ are simplified as follows:

$$\lambda = \gamma * (M - m) + m \tag{1}$$

$$\mu = \tau * (M - m) + m \tag{2}$$

So when merging sub-clusters, the spatial distance between sub-clusters can never be greater than M and not smaller than m, hence the Eq. 1. The γ factor is a hyperparameter that indicates the level of severity on the merge. If it is equal to 1, the less strict value, it tends to allow all sub-clusters to be merged and inversely if it is near to 0. Similarly, the factor τ (Eq. 2) is also a hyperparameter and has the same effect on the probabilistic distance. Equation 2 is justified by the fact that the Wasserstein distance is bounded between 0 and $(M - m)$. The lower bound is 0 because Wasserstein is a metric. The following evidence proves that the upper bound is $(M - m)$.

Proof. Given $U : \Omega \to D'$ and $V : \Omega \to D'$ two random variables with $D' \subset \mathbb{R}^+$ and Ω the set of possible outcomes. D' is the set of distances, U and V represents two sub-clusters and their inner distances. According to [14] the first Wasserstein distance between the distributions of U and V is $ws(u,v) = \int_{-\infty}^{+\infty} |F_U(u) - F_V(v)| du dv$.

F_U and F_V are respectively the Cumulative Distribution Function (CDF) of U and V. Let m be the lower bound of D' and M the upper bound of D', so $F_U(u) = F_V(v) = 0$ for $u, v \in I =]-\infty; m[$ and $F_U(u) = F_V(v) = 1$ for $u, v \in I' =]M; +\infty[$ according to CDF definition. Then,

$ws(u,v) = \int_I |F_U(u) - F_V(v)| du dv + \int_m^M |F_U(u) - F_V(v)| du dv + \int_{I'} |F_U(u) - F_V(v)| du dv = \int_I |0 - 0| du dv + \int_m^M |F_U(u) - F_V(v)| du dv + \int_{I'} |1 - 1| du dv$

$ws(u,v) = \int_m^M |F_U(u) - F_V(v)| du dv$.

By definition, $F_U : D' \to [0; 1]$ and $F_V : D' \to [0; 1]$, therefore $\int_m^M F_U(u) du \leq (M-m) \times 1$ and $\int_m^M F_V(v) dv \leq (M-m) \times 1$. As F_U and F_V are positive functions, $ws(u,v) = \int_m^M |F_U(u) - F_V(v)| du dv \leq \int_m^M F_U(u) du \leq M - m$.

3   Experiments

We conducted experiments to show the effectiveness and the robustness of DECWA compared to state-of-the-art density-based methods. It was applied to a variety of synthetic and real datasets, using distances depending on the data. First, an analysis comparing DECWA to other algorithms is performed, followed by a study of the influence of the k-nearest neighbors (k) internal parameter.

3.1   Experiment Protocol

Algorithms. DECWA was compared to DBCLASD [8], DENCLUE [9] and HDBSCAN [7]. Algorithm parameter values are defined through the repetitive application of the random search approach (1000 iterations). This, in order to obtain the best score that the four algorithms can have. DBCLASD (parameter-free and incremental) was subject to the same approach but by randomizing the order of the data points because it is sensitive to it. The hierarchical structure proposed by HDBSCAN is exploited by the Excess Of Mass (EOM) method to extract clusters, as used by the authors in [7].

Clustering quality is measured by the commonly used Adjusted Rand Index (ARI). We also report the ratio of outliers produced by each algorithm.

DECWA Environement. Several algorithms (e.g. Kruskal [15], Boruvka [16], Prim [17]) exist to generate an MST. DECWA uses Kruskal algorithm to build the MST, it has a complexity of $O(|E|log(|E|))$ with $|E|$ the cardinal of edges. Kruskal is faster than Boruvka and Prim in sparse graphs (few edges). As G is a k-nearest neighbor graph, it is often sparse because of the small value of k which is studied in Subsect. 3.2.

During the merging process, DECWA evaluates edges of G^{min} in an order which has a small influence over the result. Multiple sorting methods were tested and DECWA is not very sensitive to the order. We obtained the best results by sorting edges by weight in ascending order, therefore, we suggest this sorting method for DECWA.

To cut $p.d.f$, different methods are tested and evaluated. Inspired by Hartigan's level set idea, the minima in the $p.d.f$ curve are tested to separate distance groups (44% ARI on average). The G-means algorithm, based on Gaussian Mixture Models, was tested to separate distance groups by considering them as sub-populations of a Gaussian mixture distribution (57% ARI on average). Our mid-distance solution has better performances (67% ARI on average).

Data Sets. The results are reported in the Table 1. LD column stands for low-density, it means that the datasets have at least one low-density cluster comparing to others. SD stands for near clusters of similar densities, it means that the marked datasets have at least two overlapping clusters with similar densities. The two-dimensional, jain, compound, pathbased, and cluto-t7.10k datasets are synthetic. They contain clusters of different shapes and densities. For these data, the Euclidean distance was used.

The Iris composed of 3 classes was used with the Manhattan distance. The cardiotocography dataset is a set of fetal cardiotocogram records grouped according to 10 classes. The Canberra distance was applied. Plant dataset consists of leaves that are characterized by 64 shape measurements retrieved from its binary image. There are 100 different classes and the Euclidean distance was used. A collection of two very large biological datasets were added. GCM consists of 190 tumor samples that are represented by the expression levels of 16064 genes to diagnose 14 classes of cancer. Kidney_uterus is composed of 384 tumor samples that are expressed by 10937 genes to be classified into two classes. The Bray-Curtis distance was applied. New3, a very high-dimensional dataset, is a set of 1519 documents grouped into 6 classes. Each document is represented by the term-frequency (TF) vector of size 26833. The cosine distance was used to measure the document similarities.

Table 1. Characteristics of data sets.

Dataset	Dimensions	Size	LD	SD
jain	2	373		
compound	2	399	x	x
pathbased	2	300		x
cluto-t7.10k	2	10000		
iris	4	150		x
cardiotocography	35	2126		
plant	65	1600	x	x
GCM	16064	190	x	x
kidney_uterus	10936	384		x
new3	26833	1519		x

3.2   Results and Discussions

The results are reported in the Table 2 (best scores are in bold). In many cases, DECWA outperforms competing algorithms by a large margin on average 23%. DECWA has the best results in datasets with LD clusters with an ARI margin of 20% on average in favor of DECWA. Though datasets with high dimensions are problematic for the other algorithms, DECWA is able to give good results. There is an ARI margin of 29% on average in favor of DECWA for the three high dimensional datasets. Near clusters of similar densities are also correctly detected by DECWA. Some datasets like *iris, kidney_uterus* and *pathbased* have overlapping clusters and yet DECWA separates them correctly. There is an ARI margin of 23% in favor of DECWA for these datasets.

The outlier ratio is not relevant in case of a bad ARI score because in this case, although the points are placed in clusters, clustering is meaningless. DECWA returns the least outliers on average while having a better ARI score.

Statistical Study. We conducted a statistical study to confirm the significant difference between the algorithms and the robustness of DECWA. Firstly, a statistical test (Iman-Davenport) is performed to determine if there is a significant difference between the algorithms at the ARI level. The p-value allows us to decide on the rejection of the hypothesis of equivalence between algorithms. To reject the hypothesis, it must be lower than a significance threshold $\alpha = 0.01$. The p-value returned by the first test is $5.38e^{-5}$. It is significantly small compared to α meaning that the algorithms are different in terms of ARI performance.

Secondly, a pairwise comparison of algorithms is performed to identify the gain of one method over the others. We use the Shaffer Post-hoc test. The p-value is much lower than α when comparing DECWA to others ($7.5e^{-4}$ with DBCLASD, $4.6e^{-3}$ with DENCLUE and $3.0e^{-3}$ with HDBSCAN), which proves that DECWA is significantly different from the others. For the others, the p-value is equal to 1.0 in all the tests concerning them, which statistically shows that they are equivalent. The ranking of the algorithms according to the ARI was done by Friedman's aligned rank. DECWA is ranked as the best. Then we conclude that DECWA having the best performing is significantly different from the others.

Table 2. Experimental results

Dataset	DECWA		DBCLASD		DENCLUE		HDBSCAN	
	ARI	Outliers (%)	ARI	Outliers (%)	ARI	Outliers (%)	ARI	Outliers (%)
jain	**0.94**	**0.00**	0.90	0.03	0.45	0.06	**0.94**	0.01
compound	**0.95**	0.05	0.77	0.06	0.82	0.05	0.83	**0.04**
pathbased	**0.76**	**0.00**	0.47	0.03	0.56	**0.00**	0.42	0.02
cluto-t7.10k	**0.95**	0.03	0.79	0.06	0.34	**0.00**	**0.95**	0.10
iris	**0.77**	0.04	0.62	0.07	0.74	**0.00**	0.57	**0.00**
cardiotocography	**0.47**	0.04	0.24	0.17	**0.47**	**0.02**	0.08	0.28
plant	**0.18**	**0.03**	0.04	0.07	0.14	0.17	0.04	0.38
GCM	**0.46**	0.05	0.18	0.18	0.24	0.06	0.27	0.27
kidney_uterus	**0.80**	**0.00**	0.53	0.08	0.56	0.09	0.53	0.26
new3	**0.41**	0.01	0.09	0.45	0.08	0.13	0.13	0.27
Average	**0.67**	**0.02**	**0.46**	**0.12**	**0.44**	**0.06**	**0.48**	**0.16**

Study of the Influence of k. The MST can be used for agglomerative clustering [18], which is the case for DECWA. However, it is also pointed out that the exact distances may be lost during the process of creating the MST. A connected graph is recommended before applying an MST generation algorithm because it contains inter-sub-cluster relationships. The value of k must be large enough for the graph to be connected. The authors of [19] stipulate that for $knng$ to be connected, k is chosen to be of the order of $log(|V|)$. In the rest of this sub-section, we study the influence of k from three angles. First, we study the influence of k on the number of connected sub-graphs in the MST. Then, we evaluate the percentage of the inter-cluster distances retained by the MST that correspond to the actual two closest points belonging to different sub-clusters. Lastly, we evaluate the k influence on the results produced by DECWA. For each dataset, DECWA is called with $k = 3$, $k = log(|V|)$ and $k = \sqrt{(|V|)}$. To find h, we use the Elbow method [20] with regard to the number of peaks in order to limits the number of peaks while keeping the most significant ones which correspond the most to actual densities. γ and τ are set by using the random-search strategy. In Table 3, we report the results of the experiments in relation to the three stated angles. The numerical value of each cell in the table is the average of the results for each dataset except for diff edges which is the total percentage. $ncsb$ stands for the number of connected subgraphs. As k grows, $ncsb$ tends to decrease while the ARI increases. If $k = log(|V|)$ or $k = \sqrt{(|V|)}$, even if the connectivity is not fully ensured, DECWA remains efficient. Therefore, $k = log(|V|)$ might be a valid choice to consider rather than $k = \sqrt{|V|}$ if speed is required. Regarding the second analysis angle (diff.edges), for $k = 3$ a notable percentage of edges are different (2.3%), however it decreases as k gets bigger (0.2% for $k = log(|V|)$

and 0% for $k = \sqrt{|V|}$). Concerning the third angle, we found that the greater the k, the better the ARI. Between $k = log(|V|)$ and $k = \sqrt{(|V|)}$ the difference is only of 0.02. Regarding the outlier ratio, the difference between $k = log(|V|)$ and others is negligible.

Table 3. Influence of k from various parameterizations

k	ARI	Outliers	ncsb	diff.edges
3	0.57	0.02%	3.50	2.3%
$log(|V|)$	0.65	0.03%	1.50	0.2%
$\sqrt{(|V|)}$	0.67	0.02%	1.10	0.0%

4   Conclusion

Even if density-based clustering methods have proven to be effective for arbitrary-shaped clusters, they have difficulties to identify low-density clusters, near clusters with similar densities, and clusters in high-dimensional data. We propose DECWA, a new clustering algorithm based on spatial and probabilistic density. Our proposal consists of a new clustering algorithm based on spatial density and probabilistic approach, using a novel agglomerative method based on Wasserstein metric. We conducted an experimental study to show the effectiveness and the robustness of DECWA. For data sets with low-density clusters, near clusters and very high dimension clusters, DECWA significantly outperforms competing algorithms. Our ongoing research integrates the application of DECWA on complex data as multivariate time series.

References

1. Jain, A.: Data clustering: 50 years beyond k-means. Pattern Recogn. Lett. **31**, 651–666 (2010)
2. Kriegel, H.-P., Kröger, P., Sander, J., Zimek, A.: Density-based clustering, pp. 231–240, May 2011
3. Aggarwal, C.C., Reddy, C.K.: Data Clustering: Algorithms and Applications, 1st edn. Chapman & Hall/CRC, Boca Raton (2013)
4. Kriegel, H.-P., Kröger, P., Zimek, A.: Clustering high-dimensional data: a survey on subspace clustering, pattern-based clustering, and correlation clustering. TKDD **3**, 1–58 (2009)
5. Ester, M., Kriegel, H.-P., Sander, J., Xu, X.: A density-based algorithm for discovering clusters in large spatial databases with noise (1996)
6. Ankerst, M., Breunig, M., Kriegel, H.-P., Sander, J.: Optics: ordering points to identify the clustering structure, vol. 28, pp. 49–60, June 1999
7. Campello, R.J.G.B., Moulavi, D., Sander, J.: Density-based clustering based on hierarchical density estimates. In: Pei, J., Tseng, V.S., Cao, L., Motoda, H., Xu, G. (eds.) PAKDD 2013. LNCS (LNAI), vol. 7819, pp. 160–172. Springer, Heidelberg (2013). https://doi.org/10.1007/978-3-642-37456-2_14

8. Xu, X., Ester, M., Kriegel, H.-P., Sander, J.: A distribution-based clustering algorithm for mining in large spatial databases, pp. 324–331, January 1998

9. Hinneburg, A., Gabriel, H.-H.: DENCLUE 2.0: fast clustering based on kernel density estimation. In: R. Berthold, M., Shawe-Taylor, J., Lavrač, N. (eds.) IDA 2007. LNCS, vol. 4723, pp. 70–80. Springer, Heidelberg (2007). https://doi.org/10.1007/978-3-540-74825-0_7

10. Hartigan, J.A.: Clustering Algorithms. Wiley, New York (1975)

11. Zahn, C.T.: Graph-theoretical methods for detecting and describing gestalt clusters. IEEE Trans. Comput. **C-20**(1), 68–86 (1971)

12. Davis, R.A., Lii, K.-S., Politis, D.N.: Remarks on some nonparametric estimates of a density function. In: Selected Works of Murray Rosenblatt. SWPS, pp. 95–100. Springer, New York (2011). https://doi.org/10.1007/978-1-4419-8339-8_13

13. Villani, C.: Optimal Transport: Old and New. Springer, Heidelberg (2009). https://doi.org/10.1007/978-3-540-71050-9

14. Ramdas, A., Garcia, N., Cuturi, M.: On Wasserstein two sample testing and related families of nonparametric tests. Entropy **19**, 47 (2015)

15. Kruskal, J.B.: On the shortest spanning subtree of a graph and the traveling salesman problem. Proc. Am. Math. Soc. **7**, 48–50 (1956)

16. Nešetřil, J., Milková, E., Nešetřilová, H.: Otakar borůvka on minimum spanning tree problem translation of both the 1926 papers, comments, history. Discrete Math. **233**, 3–36 (2001)

17. Prim, R.C.: Shortest connection networks and some generalizations. Bell Syst. Tech. J. **36**(6), 1389–1401 (1957)

18. Gower, J.C., Ross, G.J.S.: Minimum spanning trees and single linkage cluster analysis. J. Royal Stat. Soc. Ser. C **18**, 54–64 (1969)

19. Brito, M., Chávez, E., Quiroz, A., Yukich, J.: Connectivity of the mutual k-nearest-neighbor graph in clustering and outlier detection. Stat. Prob. Lett. **35**, 33–42 (1997)

20. Thorndike, R.L.: Who belongs in the family? Psychometrika **18**(4), 267–276 (1953)

ODCA: An Outlier Detection Approach to Deal with Correlated Attributes

Fabrizio Angiulli, Fabio Fassetti, and Cristina Serrao[✉]

DIMES, University of Calabria, 87036 Rende, CS, Italy
{f.angiulli,f.fassetti,c.serrao}@dimes.unical.it

Abstract. Datasets from different domains usually contain data defined over a wide set of attributes or features linked through correlation relationship. Moreover, there are some applications in which not all the attributes should be treated in the same fashion as some of them can be perceived like independent variables that are responsible for the definition of the expected behaviour of the remaining ones. Following this pattern, we focus on the detection of those data objects showing an anomalous behaviour on a subset of attributes, called behavioural, w.r.t the other ones, we call contextual. As a first contribution, we exploit Mixture Models to describe the data distribution over each pair of behavioral-contextual attributes and learn the correlation laws binding the data on each bidimensional space. Then, we design a probability measure aimed at scoring subsequently observed objects based on how much their behaviour differs from the usual behavioural attribute values. Finally, we join the contributions calculated in each bidimensional space to provide a global outlierness measure. We test our method on both synthetic and real dataset to demonstrate its effectiveness when studying anomalous behaviour in a specific context and its ability in outperforming some competitive baselines.

Keywords: Anomaly detection · Mixture models · Correlation

1 Introduction

An outlier is usually perceived as an observation that deviates so much from the others in the dataset that it is reasonable to think it was generated by a different mechanism. In this paper, we start from the idea that the violation of the law according to which the points of a dataset are distributed may be a significant indication of anomaly, thus we design a strategy whose aim is to catch the relations between attributes that are lost in anomalous objects.

Consider the example reported in Fig. 1a. In this context, an obvious correlation exists between the two attributes and there could be many interpretations of what an outlier is. A method based on the spatial proximity between a point and its neighborhood will certain labels point A as an outlier, although it can be just considered as the information about an high person with a normal weight.

© Springer Nature Switzerland AG 2021
M. Golfarelli et al. (Eds.): DaWaK 2021, LNCS 12925, pp. 180–191, 2021.
https://doi.org/10.1007/978-3-030-86534-4_17

Differently, point B is closer to other data samples but it is easy to note that it violates the proportionality law. This example suggests that, to detect some subtle anomalies, the distance, in the Euclidean sense, may not be enough: objects that are distant from each other but agree with a certain law could be considered closely as their compliance with the same law makes them similar. Particularly, we propose to take into account the empirical correlations of attributes and perceive a data sample as an anomaly w.r.t a certain pair of attributes if it deviates from the expected correlation between them.

This approach is particularly suitably when a multidimensional dataset is provided and only some attributes are perceived to be directly indicative of anomaly while the other ones can have a direct effect on their expected distribution but would never be considered worth of attention for the outlier detection task.

Suppose that, within a research about obesity, you have collected information about weight, height and time spent playing sport by a group of teenagers. Clearly, neither the height nor the time spent playing sport should never be taken as direct evidence of anomaly on their own; and not even studying the correlation between them looks interesting. On the contrary, these two variables can directly affect the weight and detecting violations in their relation is interesting for the research. In such a scenario, we say that the weight is a *behavioural* attribute and call the other ones *contextual* attributes. The outlierness of an object will arise from the abnormal *behavioural* attribute in its particular context.

Moreover, it is important to point out that in real cases more than one correlation law may exists for a certain pair of attributes. Consider the example in Fig. 1b. Cloud α exhibits an evident correlation law that is violated by point A. As for cloud β, the relation between the features is weaker and points that are most likely to be reported as anomalous are those that are far from the distribution mean. Finally, cloud γ differs from the others in that it contains a few points that do not seem to comply with any of the main laws.

The measure we propose through this paper has been designed to catch all these ideas of anomaly and is strongly related to a clustering strategy able to group points in a way that takes into account the co-relations between features.

The paper is organized as follows: Sect. 2 discusses some work related with the present one; Sect. 3 formalize the problem and present the strategy we design to detect outliers; Sect. 4 presents some experimental results and Sect. 5 draws the conclusions.

2   Related Works

Outlier detection has been widely studied and some surveys are available to get in touch with this topic [7,13]. For the purposes of our discussion, we have identified some families of methods able to provide some interesting insights for the development of our approach.

A common strategy to detect anomalies is the distance-based one: the outlierness score of each data point is evaluated by taking into account the object

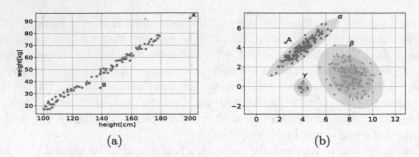

Fig. 1. Motivating example

distance from its neighborhoods assuming that outliers usually stand far from other objects [3,4,9,14]. Similarly, density-based methods compare the density of an object with the neighbors, claiming that lower density points are more likely to be outliers [6].

Another family of techniques uses a statistical point of view and exploits the deviation from the underlying generative distribution of the data [13]. Among them, in [18] the notion of outlier is captured by a strong deviation from the data dependent probabilistic distribution; instead, in [15] the data model is designed through Gaussian Mixture Models and the Mahalanobis distance [8] is used to measure the outlierness score of each object. The main drawback of these methods is related to their loss of accuracy in datasets with a large number of dimensions as a consequence of to the *curse of dimensionality* [2]. To deal with high dimensional data, subspace outlier detection methods have been designed. They rely on the projection of data into lower dimensional space [1] or focus on finding the outlying sub-spaces [10].

A line of work strongly related to our correlation-based approach is that of contextual/conditional outlier detection first introduced in [17]. In some application domains, attributes perceived by the user as potentially highly related to the outlier behavior are only a subset of the full dimensional space, while the others only provide a context for the behavior. Starting from the idea that a purely distance based approach tends to overestimate the outlierness of objects with unusual context, in [17] a statistical approach is proposed to detect outliers assuming that behavioral attributes conditionally depend on the contextual ones. Gaussian Mixture Models are employed to model both contextual and behavioral attributes and a mapping function is designed to capture their probabilistic dependence. This approach is significantly different from our in that it models data from a global point of view by learning multidimensional Mixture Models. In [12] the global expected behaviour coming from the inferred relationship between behavioral and contextual attributes from a global point of view is enriched by the concept of contextual neighborhood that take into account the object behavior within its neighborhood. Finally, in [11] an application to sensor data is presented and a robust regression model is developed to explicitly model the outliers and detect them simultaneously with the model fitting.

3   Problem Formulation

A training dataset $\mathbf{D} = \{o_1, \ldots, o_N\}$ consists of multi-set of objects defined on a set of attributes $\mathcal{A} = \{A_1, \ldots, A_M\}$ and $o_h[A_j]$ (or o_h^j for the sake of brevity) denotes the value that the object o_h assumes on the attribute A_j. A test dataset $\mathbf{T}$ is another multi-set of objects defined on the same set of attributes. The training dataset is supposed not to contains anomalies and is used to learn the model according to which the data are generated; the test dataset includes subsequently observed points whose outlierness has to be checked. A user-defined partitioning of $\mathcal{A}$ classifies the attributes into *behavioural* and *contextual*, thus we will refer to the subset of the *behavioural* attributes with $\mathcal{B} \subseteq \mathcal{A}$ and to the set of the *contextual* ones with $\mathcal{C} \subseteq \mathcal{A}$.

A *correlation pattern* between a pair of attributes (A_i, A_j), with $A_i \in \mathcal{C}$ and $A_j \in \mathcal{B}$, is a set of statistical associations that define their dependence structure.

Our goal is to assign each point in $\mathbf{T}$ with an outlierness measure depending on how much it deviates from the model that generates the data, in terms of the violation of the relational pattern that describe the expected data behaviours.

We propose a method specifically designed to solve an **O**utlier **D**etection problem in presence of **C**orrelated **A**ttribute, namely **ODCA**, that operates essentially through three steps: (I) *learning*, (II) *estimating*, (III) *combining*.

First of all, it focuses on each pair of attributes (A_i, A_j), with $A_i \in \mathcal{C}$ and $A_j \in \mathcal{B}$, and *learns* from the objects in $\mathbf{D}$ their *correlation pattern*. Then, for each bidimensional space, an *estimation* of the probability of observing each point in $\mathbf{T}$ is provided, so that a quantitative measure of its behaviour on each pair of attribute becomes available. Finally, such scores are *combined* to get an overall image of the possible outlierness level for each data point.

3.1   Learning

Let $\mathbf{D}[\{A_i, A_j\}]$ be the projection of the data set on the attribute pair (A_i, A_j). From here on we will assume that $A_i \in \mathcal{C}$ and $A_j \in \mathcal{B}$. The first step of our method consists in inferring the law according to which the projection of the dataset on such a pair is distributed. It is important to point out that, in real cases, it is difficult to note a global function according to which data is distributed, but it is more common to observe two or more data clusters; therefore, we take the perspective that the data points are generated from a mixture of distributions and we consider a Gaussian Mixture Model [5] to describe such a generation process for each bidimensional space.

A Gaussian Mixture Model can be written as a linear combination of K Gaussian each of which is described by the following parameters:

- A mean μ_k that defines its centre.
- A covariance Σ_k that defines its width.
- A mixing probability π_k that defines how big or small the Gaussian function will be, i.e. the overall probability of observing a point that comes from Gaussian k. Such coefficients must sum to 1.

If we assume that the data points are drawn independently from the distribution, then we can express the Gaussian Mixture Model in this form:

$$p(o_h^{i,j}|\mu, \Sigma, \pi) = \sum_{k=1}^{K} \pi_k \mathcal{N}(o_h^{i,j}|\mu_k, \Sigma_k) \tag{1}$$

where $o_h^{i,j}$ is the projection of the object o_h on the attributes (A_i, A_j).

Our goal is to determinate the optimal values for the parameters in Eq. (1), so we need estimate the maximum likelihood of the model by optimizing the joint probability of all object in $\mathbf{D}[\{A_i, A_j\}]$ which is defined as:

$$p(\mathbf{D}[\{A_i, A_j\}]|\mu, \Sigma, \pi) = \prod_{n=1}^{N} \sum_{k=1}^{K} \pi_k \mathcal{N}(o_n^{i,j}|\mu_k, \Sigma_k) \tag{2}$$

We optimize the logarithm of Eq. (2) by means of the iterative Expectation-Maximization (EM) algorithm. It starts by choosing some initial values for the means, covariances, and mixing coefficients. In our experimental analysis, such a choice has been performed by using the results obtained by preliminarily running a K-Means clustering. Hence, for the choice of the number of clusters K we rely on the Bayesian information criterion (BIC) [16]. The expectation (E-step) and maximization (M-step) steps are repeated until the log-likelihood function converges.

3.2   Estimating

Bidimensional Gaussian Mixure Model inferred during the learning step provides a description of the training data. If we consider each distribution in the sum on its own, we can get its principal axis $v_k^1 \in \mathbb{R}^2$, i.e. the eigenvector associated with the highest eigenvalue λ_k^1 of the covariance matrix. Moreover, each distribution has its own correlation value ρ_k that provides information about the relation between A_i and A_j and how strong this relation is. We exploit this knowledge to design our probability function that, for each distribution with a high value of ρ_k, *rewards* the points arranged along the line of greatest variance, otherwise takes into account just the distance from the distribution mean.

Consider an object $o_h \in \mathbf{T}$ and take its projection $o_h^{i,j}$ on the attribute pair (A_i, A_j). The distance of $o_h^{i,j}$ from the kth distribution in the inferred model is defined by exploiting the bidimensional Mahalanobis distance as follows:

$$D_{2D}(o_h^{i,j}, \mathcal{N}_k) = \sqrt{(o_h^{i,j} - \mu_k)^T \Sigma_k^{-1} (o_h^{i,j} - \mu_k)}$$

where μ_k and Σ_k are respectively the mean and the covariance matrix of the kth distribution, we indicate as $\mathcal{N}_k$ for the sake of simplicity. Remind that the Mahalanobis distance uniquely determinates the probability density of an observation in a normal distribution. Specifically, the square of the distance follows the chi-squared distribution with d degrees of freedom, where d is the number of dimensions of the normal distribution. Thus, when dealing with 2-dimensional distributions, the probability of observing a data point whose Mahalanobis distance is less than $D_{2D}(o_h^{i,j}, \mathcal{N}_k)$ can be expressed as:

$$\mathcal{F}_{2D}(o_h^{i,j}; \mathcal{N}_k) = 1 - e^{-D_{2D}(o_h^{i,j}, \mathcal{N}_k)^2/2} \tag{3}$$

To quantify the distance of $o_h^{i,j}$ from the direction of greatest variance we consider the value in o_h^i of the line passing through the mean μ_k of the kth distribution and whose slope is the principal component v_k^1 (see Fig. 2). In this way, we get the value the object behavioural component would have if it respected the correlation law stated by the kth distribution. Let call this value $\hat{o}_h^j$. We invoke again the Mahalanobis distance in its 1-d fashion so that the distance between $\hat{o}_h^j$ and o_h^j could be weighted by the variance $\Sigma_k^{(1,1)}$ of the kth Gaussian along the jth direction. Thus, such a distance is defined as follows:

$$D_{1D}(o_h^{i,j}, \mathcal{N}_k) = \sqrt{(o_h^j - \hat{o}_h^j)^2 / \Sigma_k^{(1,1)}}$$

As for the 2D scenario, we invoke the exponential distribution as:

$$\mathcal{F}_{1D}(o_h^j; \mathcal{N}_k) = 1 - e^{-D_{1D}(o_h^j, \mathcal{N}_k)^2/2} \tag{4}$$

Finally, we join the contribution coming from Eqs. (3) and (4) in a probability function that involves all the K components of the mixture model describing the bidimensional data distribution. Thus, the ODCA bidimensional score is defined as:

$$p_{ODCA}(o_h^{i,j} | \mathcal{N}, \pi) = 1 - \sum_{k=1}^{K} \pi_k [(1 - |\rho_k|)\mathcal{F}_{2D}(o_h^{i,j}; \mathcal{N}_k) + |\rho_k|\mathcal{F}_{1D}(o_h^j; \mathcal{N}_k)] \tag{5}$$

Function (5) considers the K components of the data model and weights each contribution through the mixing probabilities π_k. Every addend includes two parts weighted by the estimated value ρ_k. A high value for ρ_k suggests that the kth Gaussian generates strongly correlated data points; in this case the contribution coming from 1-d Mahalanobis term must be prevalent to make the function able to detect points that violate the correlation. This is the case of point A in Fig. 3. As regards to the 2-d Mahalanobis term, it will be prevalent in case of low correlation. This is what happens to point B in Fig. 3: its distance from the center of the distribution prevails over its closeness to the principal axis as the correlation is weak.

Following this definition, we are able to automatically identify the most appropriate strategy to intercept any violation and anomalous data points results to be the ones with a low probability.

Consider again the distribution of points in Fig. 1b. We use such data both as training and test set, evaluate our measure for each point and plot the distribution again in Fig. 3 choosing a chromatic shade depending on the obtained values. We run the EM algorithm with $K = 3$ and indicate the mean of each distribution with a red dot, while the vectors represent their principal components. It is important to point out that the mixture model does not provide a hard clustering for the data points, thus every distribution contributes to the evaluation of the overall measure of each point; however, the contribution of farther distributions is slight and only the points on the boundary between two or more distributions will be actually influenced by all of them.

Let focus on cloud α in Fig. 3. In this case the algorithm estimates $\rho_\alpha = 0.91$, thus its contribution will depend for the 91% from the 1-d Mahalanobis function and only for the 9% from the other. For this reason, the points arranged closer to the principal component have a lighter color than the outermost ones.

As for cloud β, the estimated correlation is $\rho_\beta = 0.28$, thus there is no strong dependency relationship and the outlierness evaluation is driven by the 2-d Mahalanobis function. Note that the influence of the distribution that generates cloud α is imperceptible in practice on cloud β as it is too far. On the contrary, there is a zone in which the lines running along the principal components of the two distributions intersect. The points in this area are correctly influenced by both distributions.

Finally, the small cloud γ has $\rho_\gamma = 0.08$. It includes very few points and their color corresponds to low probability values. This is due to mixing probabilities π_k that make each of the K components contribute to the overall probability according to the number of points for which they are responsible.

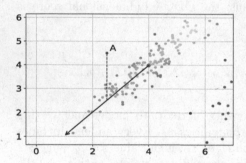

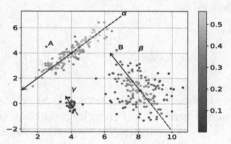

Fig. 2. 1-d Mahalanobis distance evaluation.

Fig. 3. Data points colored according to their $p_{ODCA}(\cdot)$ score.

3.3   Combining

By performing the steps discussed above we are able to represent the outlierness score of each object in **T** w.r.t any attributes pair, thus for a given *behavioural* attribute we collect a set of scores quantifying the object's outlierness in all the bidimensional spaces in which such an attribute is involved.

We summarize this information by averaging the scores within the 25th percentile to sketch the information coming from the bidimensional spaces in which the object get lower scores, that is, where it behaves in an anomalous way. In this way, we are able to provide a measure referred to each *behavioural* attribute and rank objects based on this scores. Note that, to ensure that the scores coming from different spaces become comparable, we need to standardize them before averaging.

We argue that an anomalous object will obtain low scores on most bidimensional spaces, thus the mean will confirm its anomalous nature; as for the inliers, although a bidimensional space may exists in which the probability of observing them is low, we expect that in most bidimensional space they will comply with the correlation laws, thus their $p_{ODCA}(\cdot)$ score will tend to 1.

4   Experiments

In this section we discuss the experimental results obtained by running **ODCA** on both synthetic and real datasets. Particularly, we use synthetic dataset to go in depth with the kind of outlier the method is specifically tailored, then we choose 6 real dataset and compare its performances with both state-of-art methods and contextual outlier detector.

4.1   Synthetic Dataset

The goal of the analysis performed on synthetic datasets is that of testing the effectiveness of the score definition we have provided and its ability in catching the violation of the dependence structure relying *normal* samples.

We design the training datasets by generating samples from mixtures of Gaussian distributions with specific covariance structures. Particularly, we consider models with 2 or 4 Gaussians with different configurations of the mixing probabilities π_k. For the models with two Gaussians, we first choose π_k equal to 0.5 for both the distributions and then test the configuration $\{0.75, 0.25\}$. As, for the models with 4 Gaussians we first give the same mixing parameters to all distributions and then consider the following configuration: $\{0.4, 0.3, 0.2, 0.1\}$.

To ensure that the attributes are variously correlated, we design the correlation matrix in a way that the correlation between the jth and h^{th} attribute is given by $\rho_{jh} = (-\tilde{\rho})^{|h-j|}$. Here, $\tilde{\rho}$ represents the absolute value of the highest correlation between the attribute pairs of the dataset. In the case of models with 2 Gaussian we use $\tilde{\rho} = 0.9$ for the first distribution and $\tilde{\rho} = 0.5$ for the other; in case of 4 Gaussians the values for $\tilde{\rho}$ are $\{0.9, 0.7, 0.5, 0.3\}$.

For each dataset generated in this way, we train the model on the 90% of its object and use the remaining 10% as test set. We perform each experiment 10 times by choosing different objects for the training and test set. We inject outliers by sampling the 10% of the test objects, choosing a subset of attributes and increasing or decreasing their values on such attributes of a certain quantity; some little noise is added also on the remaining attributes.

We consider a 12-attribute dataset generated according to the strategy discussed above. Because of the approach we use to generate the correlation matrix,

Table 1. Area under the ROC curve for different experimental settings. Columns refer to the parameter settings used to generate the training set, while each row concerns the attributes paired with the first one to generate the scores.

	Equal mixing prob.		No equal mixing prob.	
	n.2 Gaussians	n.4 Gaussians	n.2 Gaussians	n.4 Gaussians
3-strong	0.9907 ($\pm$0.0135)	0.9726 ($\pm$0.0470)	0.9889 ($\pm$0.0199)	0.9587 ($\pm$0.0632)
6-strong	0.9915 ($\pm$0.0125)	0.9981 ($\pm$0.0038)	1.0000 ($\pm$0.0000)	0.9764 ($\pm$0.0268)
All	0.9844 ($\pm$0.0204)	0.9989 ($\pm$0.0019)	1.0000 ($\pm$0.0000)	0.9644 ($\pm$0.0325)
6-weak	0.9840 ($\pm$0.0183)	0.9876 ($\pm$0.0171)	0.9996 ($\pm$0.0013)	0.9511 ($\pm$0.0325)
3-weak	0.9200 ($\pm$0.0624)	0.9731 ($\pm$0.0280)	0.9978 ($\pm$0.0030)	0.9476 ($\pm$0.0420)

the correlation between the first attribute and the other ones is $(-\tilde{\rho})^i$ with $i \in \{1, \ldots, 12\}$, therefore we decide to evaluate the p_{ODCA} score by first joining the contribution obtained by pairing the first attribute with the three most correlated to it (3-strong) and the three least correlated (3-weak); we then consider the six most (6-strong) and the six least (6-weak) correlated pairs, and finally all the pairs involving the first attribute.

We rank objects in the test set according to the outlierness score and evaluate the accuracy of our method in detecting outliers by the area under the ROC curve as reported in Table 1. The method looks quite robust to all the different scenarios. We argue that its strongness is the ability of taking into account, in a completely autonomous way, whether an explicit correlation law exists or not and evaluate the outlierness of each data point by the real relations binding the attribute values.

4.2   Real Dataasets

The experiments have been performed on 6 public available real dataset whose brief description is reported below. For each of them we have identified the attributes that are reasonably perceptible as *behavioral* and used the others as *contextual*.

1. **AquaticToxicity**[1] (546 × 9). The dataset reports the concentration of 8 chemicals and the LC50 index, a parameter used in environmental science to describe the toxicity of the environment. Such a parameter is used as behavioural attribute.
2. **Philips**[2] (677 × 9). For each of 677 diaphragm parts for TV sets nine characteristics were measured at the beginning of a new production line. There is no clear indication to choose an attribute as behavioural, so we analyse each

---

[1] https://archive.ics.uci.edu/ml/datasets/QSAR+aquatic+toxicity.

[2] https://www.rdocumentation.org/packages/cellWise/versions/2.2.3/topics/data_philips.

attribute as a function of all the other ones and consider the median of the accuracy we get from each experiment.

3. **Housing**[3] (20640 × 9). It includes information on house prices in California w.r.t some environmental characteristics. The price has been used as behavioural attribute.

4. **Fish**[4] (908 × 7). The dataset reports the concentration of 6 chemicals and a parameter describing the incidence in fish mortality. Such a parameter is used as behavioural attribute.

5. **Boston**[5] (506 × 13). The data concerns housing values in suburbs of Boston. We remove the binary variable about closeness to Charles River and choose the Median value of owner-occupied homes as behavioural and the other economic data as contextual.

6. **BodyFat**[6] (252 × 15). The data contain estimates of the percentage of body fat determined by underwater weighing and various body circumference measurements for a group of people. The attribute on body percentage has been used as behavioural and other physical features as contextual.

Each dataset has been analysed by each by our techniques and the competitor ones using the same 10-fold cross-validation. For each fold, we divide the dataset into ten blocks of near-equal size and, for every block in turn, each method has been trained on the remaining blocks and tested on the hold-out one.

None of these datasets is equipped with a ground truth, so we inject outliers within the test set. With this aim, we first calculate the outlierness scores of the points in the hold-out block before the perturbation, then we randomly select the 10% of the objects resulting to be inliers, i.e. whose scores is greater than the median of all the scores, and increase or decrease their values on the *behavioural* attributes; finally some little noise is added also on the *contextual* attributes.

As for the baseline methods to perform the comparative analysis, we consider both state-of-art outlier detectors and methods specifically designed to detect outliers complying to our definition; within this category we select ROCOD [12], and consider both its linear and no-linear approach as it seems to be the most competitive w.r.t the other baseline methods we have revised from literature. Thus, we test the following techniques:

1. **KNN** [3]. Each point from the test set is scored according to the distance from its kth neighbor among the points in the training set (we use $k = 10$ in our experiments). A larger distance indicates a more anomalous point.

2. **LOF** [6]. The Local Outlier Factor of each test point is computed with respect to the training dataset.

3. **GMM** [18]. A multidimensional Gaussian Mixture Distribution is used to fit the data from the training set. The inferred density function is used to score test set points and low-density points are perceived as outliers.

---

[3] https://www.kaggle.com/camnugent/california-housing-prices.
[4] https://archive.ics.uci.edu/ml/datasets/QSAR+fish+toxicity.
[5] https://archive.ics.uci.edu/ml/machine-learning-databases/housing/.
[6] http://staff.pubhealth.ku.dk/~tag/Teaching/share/data/Bodyfat.html.

4. **ROCOD-L** [12]. An ensemble of local and global expected behaviour is used to detect outliers. The local expected behavior of a test data point is computed by averaging the values of the behavioural attributes of the subset of training data point sharing a similar values on contextual attributes with the test point. The global expected behaviour is modelled through linear ridge regression.
5. **ROCOD-NL** [12]. A non-linear version of ROCOD which use tree regression to model the global expected behaviour (Fig. 4).

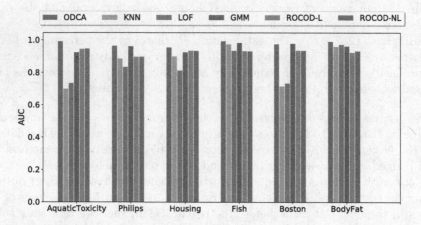

Fig. 4. Accuracy bar plot. It reports the area under the ROC curve each method gets on different dataset.

Traditional outlier detector like KNN and LOF are good in identifying those objects that fall in low density areas as a result of the perturbation protocol. However, their top ranked objects include also those data points that are extreme on *contextual* attributes without violating any correlation law. Hence for GMM, we argue that the increase in accuracy of our method is due to its ability of going in depth with the point's behaviour on each bidimensional space. Finally, ROCOD, both in its linear and non-linear version, do not take into account the presence of a mixture of laws describing the relation between *behavioural* and *contextual* attributes, causing a loss in accuracy.

5   Conclusions

We have proposed an outlier detection method that attempts to detect outliers based on empirical correlations of attributes. The method analyses the data and learns which values are usual or typical for the behavioural attributes, so that when a subsequent data point is observed it is labeled as anomalous or not depending on how much its behaviour differs from the usual behavioural attribute values. As the correlation is well defined only between pairs of attributes we take the perspective of studying each behavioural attribute paired with a contextual one at a time. Moreover, the use of mixture models allow us to detect complex correlation patterns. We move to a global viewpoint by joining the bidimensional scores, however the combining strategy could be involved in future researches aimed at improving the characterization of the outlier object behaviour.

References

1. Aggarwal, C.C., Yu, P.S.: Outlier detection for high dimensional data. In: Proceedings of the 2001 ACM SIGMOD MOD Conference, pp. 37–46 (2001)
2. Angiulli, F.: On the behavior of intrinsically high-dimensional spaces: distances, direct and reverse nearest neighbors, and hubness. J. Mach. Learn. Res. **18**, 170:1–170:60 (2017)
3. Angiulli, F., Basta, S., Pizzuti, C.: Distance-based detection and prediction of outliers. IEEE TKDE **18**(2), 145–160 (2005)
4. Angiulli, F., Fassetti, F.: DOLPHIN: an efficient algorithm for mining distance-based outliers in very large datasets. ACM TKDD **3**(1), 1–57 (2009)
5. Bishop, C.M.: Pattern Recognition and Machine Learning. Springer, New York (2006)
6. Breunig, M.M., Kriegel, H.-P., Ng, R.T., Sander, J.: LOF: identifying density-based local outliers. In: Proceedings of the 2000 ACM SIGMOD International Conference on Management of Data, pp. 93–104 (2000)
7. Chandola, V., Banerjee, A., Kumar, V.: Anomaly detection: a survey. ACM Comput. Surv. (CSUR) **41**(3), 1–58 (2009)
8. Mahalanobis, P.C., et al.: On the generalized distance in statistics. In: Proceedings of the National Institute of Sciences of India, vol. 2 (1936)
9. Knorr, E.M., Ng, R.T.: Algorithms for mining distance-based outliers in large datasets. In: VLDB, vol. 98, pp. 392–403. Citeseer (1998)
10. Kriegel, H.-P., Kröger, P., Schubert, E., Zimek, A.: Outlier detection in arbitrarily oriented subspaces. In: 2012 IEEE 12th ICDM, pp. 379–388 (2012)
11. Kuo, Y.-H., Li, Z., Kifer, D.: Detecting outliers in data with correlated measures. In: Proceedings of the 27th CIKM, pp. 287–296 (2018)
12. Liang, J., Parthasarathy, S., Robust contextual outlier detection: where context meets sparsity. In: Proceedings of the 25th ACM CIKM, pp. 2167–2172 (2016)
13. Markou, M., Singh, S.: Novelty detection: a review-part 1: statistical approaches. Sig. Process. **83**(12), 2481–2497 (2003)
14. Ramaswamy, S., Rastogi, R., Shim, K.: Efficient algorithms for mining outliers from large data sets. In: Proceedings of the 2000 ACM SIGMOD International Conference on Management of Data, pp. 427–438 (2000)
15. Roberts, S., Tarassenko, L.: A probabilistic resource allocating network for novelty detection. Neural Comput. **6**(2), 270–284 (1994)
16. Schwarz, G., et al.: Estimating the dimension of a model. Ann. Stat. **6**(2), 461–464 (1978)
17. Song, X., Mingxi, W., Jermaine, C., Ranka, S.: Conditional anomaly detection. IEEE TKDE **19**(5), 631–645 (2007)
18. Yamanishi, K., Takeuchi, J.-I., Williams, G., Milne, P.: On-line unsupervised outlier detection using finite mixtures with discounting learning algorithms. Data Min. Knowl. Discov. **8**(3), 275–300 (2004). https://doi.org/10.1023/B:DAMI.0000023676.72185.7c

A Novel Neurofuzzy Approach for Semantic Similarity Measurement

Jorge Martinez-Gil[1]([✉]), Riad Mokadem[2], Josef Küng[3],
and Abdelkader Hameurlain[2]

[1] Software Competence Center Hagenberg GmbH, Softwarepark 21,
4232 Hagenberg, Austria
`jorge.martinez-gil@scch.at`
[2] IRIT Toulouse, Route de Narbonne 118, 31062 Toulouse Cedex, France
[3] Johannes Kepler University Linz, Altenbergerstraße 69, 4040 Linz, Austria

Abstract. The problem of identifying the degree of semantic similarity
between two textual statements automatically has grown in importance
in recent times. Its impact on various computer-related domains and
recent breakthroughs in neural computation has increased the opportunities for better solutions to be developed. This research takes the research
efforts a step further by designing and developing a novel neurofuzzy approach for semantic textual similarity that uses neural networks and fuzzy
logics. The fundamental notion is to combine the remarkable capabilities
of the current neural models for working with text with the possibilities
that fuzzy logic provides for aggregating numerical information in a tailored manner. The results of our experiments suggest that this approach
is capable of accurately determining semantic textual similarity.

Keywords: Data integration · Neurofuzzy · Semantic similarity ·
Deep learning applications

1 Introduction

Data mining and knowledge discovery techniques have long been trying to boost
the decision-making capabilities of human experts in a wide variety of academic
disciplines and application scenarios. In addition, these techniques have greatly
facilitated the development of a new generation of computer systems. These
systems are designed to solve complex problems using expert-generated knowledge rather than executing standard source code. However, this notion has been
evolving towards more effective and efficient models and offers more flexibility in
supporting the human judgment in decision-making. In this work, we look at the
field of semantic textual similarity. That is the possibility of boosting the capability of a human expert in deciding whether two pieces of textual information
could be considered similar. In this way, automatic data integration techniques
can be notably improved.

© Springer Nature Switzerland AG 2021
M. Golfarelli et al. (Eds.): DaWaK 2021, LNCS 12925, pp. 192–203, 2021.
https://doi.org/10.1007/978-3-030-86534-4_18

Over the last two decades, different approaches have been put forward for computing semantic similarity using a variety of methods and techniques [11]. In this way, there are already many solutions for automatically calculating the semantic similarity between words, sentences, and even documents. Currently, the solutions that can obtain the best results are those based on neural networks such as USE [4], BERT [9] or ELMo [25]. However, there is still much room for improvement since the development of these models is still in its infancy. Many of the functions they implement are trivial, and it is to be expected that as more sophisticated approaches are investigated, the results to be achieved could be better.

For these reasons, we have focused on a novel approach that is slowly making its way into the literature: neurofuzzy systems. Systems of this kind are built by a clever combination of artificial neural networks and fuzzy logic. These systems attract much attention because they can bring together the significant advantages of both worlds [27]. However, its application in the domain of semantic similarity remains unexplored.

We want to go a step further to design a new neurofuzzy approach that might be able to determine automatically and with high accuracy the degree of semantic similarity between pieces of textual information. To do that, we propose to follow a concurrent fuzzy inference neural network (FINN) approach being able to couple the state-of-the-art models from the neural side together with the state-of-the-art from the fuzzy side. This approach is expected to achieve highly accurate results as it brings together the computational power of neural networks with the capability of information fusion from fuzzy logics. Thus, the contributions of this research work can be summarized as follows:

- We propose for the first time a neurofuzzy schema for semantic similarity computation that combines the ability of neural networks to transform pieces of textual information into vector information suitable for processing by automatic methods with the advantages of personalized aggregation and decoding offered by fuzzy logics.
- We have subjected our proposal to an empirical study in which we compare it with state-of-the-art solutions in this field. The results obtained seem to indicate that our proposal yields promising results.

The rest of this paper is structured as follows: in Sect. 2, we present the state-of-the-art concerning the automatic computation of semantic similarity when working with textual information and recent solutions based on neurofuzzy systems to solve practical application problems. In Sect. 3, we provide the technical explanation on which our neurofuzzy approach is based. In Sect. 4, we undertake an empirical study that compares our approach with those that make up the state-of-the-art. Finally, we highlight the lessons that can be drawn from the present work and point out possible future lines of research.

2   State-of-the-Art

The semantic similarity field attracts much attention because it represents one of the fundamental challenges that can advance several fields and academic disciplines [17]. The possibility that a computer can automatically determine the degree of similarity between different pieces of textual information regardless of their lexicography can be very relevant. This means that areas such as data integration, question answering, or query expansion could greatly benefit from any progress in this area.

To date, numerous solutions have been developed in this regard. These solutions range from traditional techniques using manually compiled synonym dictionaries such as Wordnet [15], to methods using the web as a large corpus such as Normalized Google Distance [5] through the classical taxonomy-based techniques [26] or the ones based on corpus statistics [13].

Besides, more and more solutions have been developed that are valid in a wide range of domains of different nature. Some of these solutions are based on the aggregation of atomic methods to benefit from many years of research and development in semantic similarity measures [21]. Moreover, there have been breakthroughs that have completely revolutionized the field of semantic similarity. One of the most promising approaches has been word embeddings [23]. Where solutions of a neural nature have been able to reduce pieces of text to feature vectors of a numerical nature so that they are much more suitable for automatic processing by computers [14]. These approximations are so robust that they can determine the degree of similarity of cross-lingual expressions [10].

Neurofuzzy systems have begun to be used in many application domains due to the versatility they offer. Neurofuzzy systems have the great advantage of combining the human-like reasoning of fuzzy systems through a linguistic model based on IF-THEN rules with the tremendous computational power to discover patterns of neural networks. Neurofuzzy systems are considered universal approximators because of their ability to approximate any mathematical function, which allows them to be highly qualified for problems related to automatic learning.

Concerning these neurofuzzy systems' neural side, some neural approaches have been recurrently used to work with text—for example, automatic translation or text auto-completion. The problem is that these models are usually not very good at capturing long-term dependencies. For this reason, transformer architectures have recently emerged [9]. This kind of architecture uses a particular type of attention known as self-attention [8].

However, all these architectures of neural nature that have been presented have only used simplistic ways of aggregation and decoding of the last neural layer to date. The operations that can be found recurrently in the literature are cosine similarity, manhattan distance, euclidean distance, or inner product. This is where our contribution comes into play since we propose a fuzzy logic-based solution that can model a much more sophisticated interaction between the numerical feature vectors generated by the neural part.

3   A Novel Neurofuzzy Approach for Semantic Similarity

Fuzzy logic can offer computational methods that aim to formalize reasoning methods that are considered approximate. We are here considering Mamdani fuzzy inference [19] since it is considered an optimal method for developing control systems governed by a set of rules very close to natural language. In Mamdani fuzzy inference systems, the output of each rule is always a fuzzy set. Since systems of this kind have a rule base that is very intuitive and close to natural language, they are often used in expert systems to model the knowledge of human experts.

The reason to use fuzzy logics is that rules can also be derived analytically when it is impossible to count on the expert's help, as is the case in our approach. In our specific scenario, we use aggregation controllers. These controllers are usually divided into several components including a database of terms such as $\mu_{\widetilde{S}}(x)$ that states the membership of x in $\widetilde{S} = \left\{ \int \frac{\mu_{\widetilde{S}}(x)}{x} \right\}$ what is usually defined as $\mu_{\widetilde{S}}(x) \in [0,1]$, and a non-empty set of rules. In this way, the terms associated with the database can be used to characterize the rules.

Moreover, the input values need to be encoded according to the terms from the controller, so that $\widetilde{I} = \mu_1 Q(x_1) + \mu_2 Q(x_2) + ... + \mu_n Q(x_n)$, whereby μ_i is the term associated with the transformation of x_i into the set $Q(x_i)$.

Last but not least, we need to define the terms on the basis of membership functions so that: $\widetilde{T} = \left\{ (x, \mu_{\widetilde{T}}(x)) \mid x \in U \right\}$. The great advantage of this approach is that a wide range of membership functions can be defined by just using a limited number of points which represents an advantage for us when coding possible solutions in the form of individuals from an evolving population.

Working with Mamdani fuzzy systems [19] also means that the result of the inference will be a set such as $\widetilde{O} = \left\{ \int \frac{\mu_{\widetilde{O}}(v)}{v} \right\}$. Therefore, the output might be a real value representing the result of aggregating the input values. One of the traditional advantages of Mandami's models concerning other approaches, e.g., Tagaki-Sugeno's [28], is that they facilitate interpretability. This is because the Mamdani inference is well suited to human input while the Tagaki-Sugeno inference is well suited to analysis [7].

On the other hand, the neural side will use transformers that are suitable models for translations between abstract representations. The transformer models consist of an encoder-decoder architecture. It is necessary to feed the encoder with the input textual information. From there, the encoder learns to represent the input information and sends this representation to the decoder. The decoder receives the representation and generates the output information to be presented to the user. The way of working is like this: for each attention item, the transformer model learns three weight matrices; the query matrix W_Q, the key matrix W_K, and the value matrix W_V. For each token i, the input embedding x_i is multiplied with each of the three matrices to produce a query array $q_i = x_i W_Q$, a key array $k_i = x_i W_K$, and a value array $v_i = x_i W_V$. Attention weights are calculated using the query and key arrays in the following way: the attention weight a_{ij} from token i to token j is the product between q_i and k_j. The attention weights are divided by the square root of the key arrays' dimension, $\sqrt{d_k}$, which stabilizes gradients during training, and it is processed by a function $softmax$, which normalizes the weights to sum to 1.

Figure 1 shows the mode of operation of a concurrent FINN architecture. The input data is first processed by the neural network (i.e., transformer), which is very good with the vectorization of the pieces of textual information into numerical feature vectors. These feature vectors will correspond to the membership functions of the fuzzy module at a later stage. The coupling of the modules of neural nature with that of fuzzy logic nature should obtain good results once a training phase has correctly calibrated all their parameters. This architecture is more complex and more powerful than the more advanced neural models because it adds the last layer that can process the results in a much more sophisticated way.

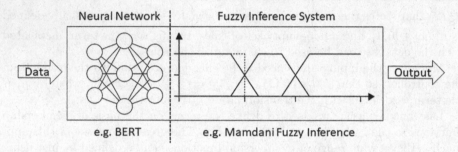

Fig. 1. Architecture of our neurofuzzy approach. The neural and fuzzy models are coupled together and then trained in order to be properly calibrated to solve the scenario that we wish

The appropriate combination of the two approaches gives rise to a concurrent FINN, which we intend to obtain good results with when trying to automatically determine the degree of semantic similarity between two textual expressions that are analogous but have been represented using different lexicographies. Also, another great advantage of these systems is that they can be trained separately or together. This allows them to benefit from great flexibility and versatility. For example, not everyone needs to develop a neural solution from scratch because the existing ones are of high quality and have been trained on hard-to-access corpora.

Finally, it is worth mentioning that our learning process is guided by an evolutionary strategy that tries to find the best parameters in the neurofuzzy model, although trying to avoid an over-fitting situation. We have opted for the classical solution for an elitist evolution model with a low mutation rate, which allows automatically exploring of the solution space, although it also allows a low rate of random jumps in search of better solutions [1]. The following pseudocode shows the rationale behind this approach.

The evolutionary strategy allows us to optimally calibrate the following parameters: the transformer model to be used, how the operations in the last

---

Algorithm 1. Pseudo-code for the evolutionary strategy to obtain optimal FINN model

```
 1: procedure CALCULATION OF THE BEST POSSIBLE FINN MODEL
 2:      RandomIndividuals (population)
 3:      calculateFitness (population)
 4:      while (NOT stop condition) do
 5:         for (each individual)
 6:            parents ← selectionOfIndividuals ()
 7:            offspring ← binCrossOver (parents)
 8:            offspring ← randomMutation (offspring)
 9:            calculateFitness (offspring)
10:            population ← updatePopulation (offspring)
11:         endfor
12:      endwhile
13:      return Model(population)
```

---

layer of the neural network will be computed, the fuzzy sets and membership functions, the IF-THEN rules that best fit the input data, as well as the defuzzification method.

4 Experimental Study

This section describes our strategy's experimental setup, including the benchmark dataset that we have used and the evaluation criteria that we are following, and the configuration of the considered methods. After that, we perform an exhaustive analysis of the different approaches considered and the empirical results we have achieved. Finally, we offer a discussion of the results we have obtained.

4.1 Datasets and Evaluation Criteria

We have used the most widely used general-purpose benchmark dataset in this field to carry out our experiments. Our approach's behavior concerning this dataset will give us an idea of how our approach works. This benchmark dataset is the so-called MC30, or Miller & Charles dataset [24] that consists of 30-word pairs that everyone might use.

Evaluating the techniques to discover semantic similarity using correlation coefficients can be done in two different ways. First, it is possible to use Pearson's correlation coefficient, which can measure the degree of correlation between the background truth and the machine results, whereby a and b are respectively the source and target pieces whose degree of semantic similarity is to be compared

$$\sigma = \frac{n \sum a_i b_i - \sum a_i \sum b_i}{\sqrt{n \sum a_i^2 - (\sum a_i)^2} \sqrt{n \sum b_i^2 - (\sum b_i)^2}} \qquad (1)$$

The second way to proceed is the so-called, Spearman Rank Correlation, whereby the aim is to measure the relative order of the results provided by the technique.

$$\rho = 1 - \frac{6 \sum d_i^2}{n(n^2 - 1)} \tag{2}$$

Being $d_i = rg(X_i) - rg(Y_i)$ the difference between the two ranks of each array, and n is the size of both arrays.

The significant difference between the two correlation methods is that while Pearson is much better at determining the total order of the dataset results, Spearman is more suitable for determining a partial order.

4.2   Configuration

As we have already mentioned, the training phase is performed by an evolutionary learning strategy. Therefore, we have had to perform a grid search to determine which are the best parameters for that strategy. Since the search space is really huge, we had to narrow down the search intervals. In this way, the identified parameters of our evolutionary strategy are the following:

- Representation of genes (binary, real): **real**
- Population size [10, 100]: **42**
- Crossover probability [0.3, 0.95]: **0.51**
- Mutation probability [0.01, 0.3]: **0.09**
- Stop condition: Iterate over (1,000–100,000): **100,000** generations

In addition, evolutionary learning techniques are non-deterministic in that they rely on randomness components to generate initial populations and search for model improvements through small mutation rates (9% in our case). Therefore, the results reported are always the average of several independent runs, as we will explain later.

It is necessary to note that the hardware used has been an Intel Core i7-8700 with CPU 3.20 GHz and 32 GB of RAM over Windows 10 Pro. Furthermore, we rely on the implementation of Cingolani's fuzzy engine [6] as well as the implementations of USE [4], BERT [9], or ELMo [25] that their authors have published initially. In that sense, training using different text corpora may indeed yield different results. However, such a study is outside the scope of this work and would be interesting future work. Finally, it is necessary to remark that the average and maximum training times are reported in the following section.

4.3   Results

In this section, we show the empirical results that we have obtained. In Table 1, it is possible to see a summary of the results we have obtained when solving the MC30 benchmark dataset using the Pearson correlation coefficient. This means that we are looking for the capability of the different solutions to establish a total order. Please note that the results reported for our approach are based on

Table 1. Results over the MC30 dataset using Pearson Correlation

Approach	Score	p-value
Google distance [5]	0.470	$8.8 \cdot 10^{-3}$
Huang et al. [12]	0.659	$7.5 \cdot 10^{-5}$
Jiang & Conrath [13]	0.669	$5.3 \cdot 10^{-5}$
Resnik [26]	0.780	$1.9 \cdot 10^{-7}$
Leacock & Chodorow [16]	0.807	$4.0 \cdot 10^{-8}$
Lin [18]	0.810	$3.0 \cdot 10^{-8}$
Faruqui & Dyer [10]	0.817	$2.0 \cdot 10^{-8}$
Mikolov et al. [23]	0.820	$2.2 \cdot 10^{-8}$
CoTO [20]	0.850	$1.0 \cdot 10^{-8}$
FLC [22]	0.855	$1.0 \cdot 10^{-8}$
Neurofuzzy (median)	0.861	$4.9 \cdot 10^{-9}$
Neurofuzzy (maximum)	0.867	$1.0 \cdot 10^{-9}$

ten independent executions due to the non-deterministic nature of the learning strategy. So we report the median value and the maximum value achieved.

Table 2 shows the results obtained for Spearman's correlation coefficient. This means that we evaluate the capability of the existing approaches when determining a partial order between the cases of the MC30 dataset. Once again, we report the median value and the maximum value achieved.

Table 2. Results over the MC30 dataset using Spearman Rank Correlation

Approach	Score	p-value
Jiang & Conrath [13]	0.588	$8.8 \cdot 10^{-3}$
Lin [18]	0.619	$1.6 \cdot 10^{-4}$
Aouicha et al. [2]	0.640	$8.0 \cdot 10^{-5}$
Resnik [26]	0.757	$5.3 \cdot 10^{-7}$
Mikolov et al. [23]	0.770	$2.6 \cdot 10^{-7}$
Leacock & Chodorow [16]	0.789	$8.1 \cdot 10^{-8}$
Bojanowski et al. [3]	0.846	$1.1 \cdot 10^{-9}$
Neurofuzzy (median)	0.851	$1.0 \cdot 10^{-9}$
Neurofuzzy (maximum)	0.868	$4.6 \cdot 10^{-9}$
FLC [22]	0.891	$1.0 \cdot 10^{-10}$

Finally, we offer a study of the time it takes to train our neurofuzzy systems. Figure 2 shows the training that has been performed to obtain the Pearson correlation coefficient. As we are working with non-deterministic methods, the results presented are equivalent to an average of ten independent runs of which we plot the minimum (red), the median (blue) and the maximum (black).

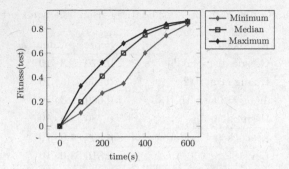

Fig. 2. Fitness evolution for the Pearson correlation. The red, blue, and black plots represent the worst, median and best cases respectively (Color figure online)

While Fig. 3 shows the time taken to properly set up our approach to solve the Spearman Rank Correlation. As in the previous case, the results we can see in the plot are the result of ten independent runs of which we plot the minimum, the median and the maximum values that we have achieved.

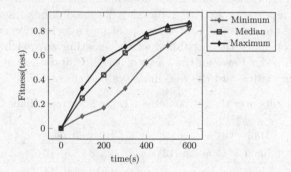

Fig. 3. Fitness evolution for the Spearman Rank correlation. The red, blue, and black plots represent the worst, median and best cases respectively (Color figure online)

As it can be seen, our approach presents several significant advantages concerning the works that make up the state-of-the-art. As for techniques based on neurofuzzy hybridization, no work has yet been done in this area to the best of our knowledge. However, our previous experience designing solutions based on fuzzy logic led us to think that combining the human-like reasoning of fuzzy logics with neural networks' learning capability would yield quite good results. Training a neurofuzzy system indeed involves a significant consumption of resources in the form of time, but it is also true that once trained, the results are pretty good. Moreover, the model can be reused. Even mature transfer learning techniques can facilitate its application in analog-nature problems in different scenarios.

5   Conclusions and Future Work

We have presented a novel approach for the automatic computation of the degree of similarity between textual information pieces. Our approach is novel because it is the first time that a neurofuzzy system is proposed to deal with the problem. We think that a neurofuzzy system is appropriate in this situation since it can combine the high capabilities of neural nature solutions to extract and convert features associated with text expressed in natural language with the ability of fuzzy logics to aggregate and decode in a personalized way information of numerical nature.

The results we have obtained show that our approach can achieve results in line with the state-of-the-art, even without being specifically trained. These promising results rely on solutions of neural nature whose accuracy is highly contrasted together with fuzzy logic system, which has a great capacity to aggregate intermediate results and decode them into the values for which it has been trained.

Besides, the two parts that make up our system can be trained separately. For example, we can use a highly constrained accuracy model such as BERT combined with a classical Mamdani inference model, which usually gives outstanding results. In this way, our system is built based on building blocks that give it flexibility and versatility not known so far in semantic similarity measurement.

As future work, we plan to explore other approaches to assess similarity automatically. We have worked here with monolingual semantic similarity, i.e., all pieces of textual information were expressed in English. However, a pending issue is to study the problem from a cross-lingual perspective. The other pending issue would be how to improve the interpretability of the resulting system. Due to the black-box model that is implemented in the neural part, a solution must be found so that people can understand this model from the beginning to the end.

Acknowledgements. This work has been supported by the Austrian Ministry for Transport, Innovation and Technology, the Federal Ministry of Science, Research and Economy, and the State of Upper Austria in the frame of the COMET center SCCH. By the project FR06/2020 by International Cooperation & Mobility (ICM) of the Austrian Agency for International Cooperation in Education and Research (OeAD-GmbH). We would also thank 'the French Ministry of Foreign and European Affairs' and 'The French Ministry of Higher Education and Research' which support the Amadeus program 2020 (French-Austrian Hubert Curien Partnership – PHC) Project Number 44086TD.

References

1. Angelov, P.P., Buswell, R.A.: Automatic generation of fuzzy rule-based models from data by genetic algorithms. Inf. Sci. **150**(1–2), 17–31 (2003)
2. Aouicha, M.B., Taieb, M.A.H., Hamadou, A.B.: LWCR: multi-layered Wikipedia representation for computing word relatedness. Neurocomputing **216**, 816–843 (2016)
3. Bojanowski, P., Grave, E., Joulin, A., Mikolov, T.: Enriching word vectors with subword information. Trans. Assoc. Comput. Linguistics **5**, 135–146 (2017)
4. Cer, D., et al.: Universal sentence encoder for English. In: Blanco, E., Lu, W. (eds.) Proceedings of the 2018 Conference on Empirical Methods in Natural Language Processing, EMNLP 2018: System Demonstrations, Brussels, Belgium, 31 October–4 November 2018, pp. 169–174. Association for Computational Linguistics (2018)
5. Cilibrasi, R., Vitányi, P.M.B.: The google similarity distance. IEEE Trans. Knowl. Data Eng. **19**(3), 370–383 (2007)
6. Cingolani, P., Alcalá-Fdez, J.: jFuzzyLogic: a java library to design fuzzy logic controllers according to the standard for fuzzy control programming. Int. J. Comput. Intell. Syst. **6**(sup1), 61–75 (2013)
7. Cordón, O.: A historical review of evolutionary learning methods for Mamdani-type fuzzy rule-based systems: designing interpretable genetic fuzzy systems. Int. J. Approx. Reason. **52**(6), 894–913 (2011)
8. Dai, B., Li, J., Xu, R.: Multiple positional self-attention network for text classification. In: The Thirty-Fourth AAAI Conference on Artificial Intelligence, AAAI 2020, The Thirty-Second Innovative Applications of Artificial Intelligence Conference, IAAI 2020, The Tenth AAAI Symposium on Educational Advances in Artificial Intelligence, EAAI 2020, New York , NY, USA, 7–12 February 2020, pp. 7610–7617. AAAI Press (2020)
9. Devlin, J., Chang, M.W., Lee, K., Toutanova, K.: Bert: pre-training of deep bidirectional transformers for language understanding. arXiv preprint arXiv:1810.04805 (2018)
10. Faruqui, M., Dyer, C.: Improving vector space word representations using multilingual correlation. In: Proceedings of the 14th Conference of the European Chapter of the Association for Computational Linguistics, EACL 2014, Gothenburg, Sweden, 26–30 April 2014, pp. 462–471 (2014)
11. Harispe, S., Ranwez, S., Janaqi, S., Montmain, J.: Semantic Similarity from Natural Language and Ontology Analysis. Synthesis Lectures on Human Language Technologies. Morgan & Claypool Publishers (2015)
12. Huang, E.H., Socher, R., Manning, C.D., Ng, A.Y.: Improving word representations via global context and multiple word prototypes. In: The 50th Annual Meeting of the Association for Computational Linguistics, Proceedings of the Conference, Jeju Island, Korea, 8–14 July 2012, Volume 1: Long Papers, pp. 873–882 (2012)
13. Jiang, J.J., Conrath, D.W.: Semantic similarity based on corpus statistics and lexical taxonomy. In: Proceedings of the 10th Research on Computational Linguistics International Conference, ROCLING 1997, Taipei, Taiwan, August 1997, pp. 19–33 (1997)
14. Lastra-Díaz, J.J., García-Serrano, A., Batet, M., Fernández, M., Chirigati, F.: HESML: a scalable ontology-based semantic similarity measures library with a set of reproducible experiments and a replication dataset. Inf. Syst. **66**, 97–118 (2017)

15. Lastra-Díaz, J.J., Goikoetxea, J., Taieb, M.A.H., García-Serrano, A., Aouicha, M.B., Agirre, E.: A reproducible survey on word embeddings and ontology-based methods for word similarity: linear combinations outperform the state of the art Eng. Appl. Artif. Intell. **85**, 645–665 (2019)
16. Leacock, C., Chodorow, M.: Combining local context and wordnet similarity for word sense identification. WordNet Electron. Lexical Database **49**(2), 265–283 (1998)
17. Li, Y., Bandar, Z., McLean, D.: An approach for measuring semantic similarity between words using multiple information sources. IEEE Trans. Knowl. Data Eng. **15**(4), 871–882 (2003)
18. Lin, D.: An information-theoretic definition of similarity. In: Proceedings of the Fifteenth International Conference on Machine Learning (ICML 1998), Madison, Wisconsin, USA, 24–27 July 1998, pp. 296–304 (1998)
19. Mamdani, E.H., Assilian, S.: An experiment in linguistic synthesis with a fuzzy logic controller. Int. J. Hum.-Comput. Stud. **51**(2), 135–147 (1999)
20. Martinez-Gil, J.: CoTO: a novel approach for fuzzy aggregation of semantic similarity measures. Cogn. Syst. Res. **40**, 8–17 (2016)
21. Martinez-Gil, J.: Semantic similarity aggregators for very short textual expressions: a case study on landmarks and points of interest. J. Intell. Inf. Syst. **53**(2), 361–380 (2019). https://doi.org/10.1007/s10844-019-00561-0
22. Martinez-Gil, J., Chaves-González, J.M.: Automatic design of semantic similarity controllers based on fuzzy logics. Expert Syst. Appl. **131**, 45–59 (2019)
23. Mikolov, T., Sutskever, I., Chen, K., Corrado, G.S., Dean, J.: Distributed representations of words and phrases and their compositionality. In: Advances in Neural Information Processing Systems 26: 27th Annual Conference on Neural Information Processing Systems 2013. Proceedings of a meeting held 5–8 December 2013, Lake Tahoe, Nevada, United States, pp. 3111–3119 (2013)
24. Miller, G., Charles, W.: Contextual correlates of semantic similarity. Lang. Cogn. Process. **6**(1), 1–28 (1991)
25. Peters, M.E., et al.: Deep contextualized word representations. In: Proceedings of NAACL-HLT, pp. 2227–2237 (2018)
26. Resnik, P.: Semantic similarity in a taxonomy: an information-based measure and its application to problems of ambiguity in natural language. J. Artif. Intell. Res. **11**, 95–130 (1999)
27. Rutkowski, L., Cpalka, K.: Flexible neuro-fuzzy systems. IEEE Trans. Neural Networks **14**(3), 554–574 (2003)
28. Takagi, T., Sugeno, M.: Fuzzy identification of systems and its applications to modeling and control. IEEE Trans. Syst. Man Cybern. **15**(1), 116–132 (1985)

Data Warehouse Processes and Maintenance

Integrated Process Data
and Organizational Data Analysis
for Business Process Improvement

Alexis Artus, Andrés Borges, Daniel Calegari[⊠], and Andrea Delgado

Instituto de Computación, Facultad de Ingeniería, Universidad de la República,
Montevideo, Uruguay
{alexis.artus,juan.borges,dcalegar,adelgado}@fing.edu.uy

Abstract. Neither a compartmentalized vision nor coupling of process data and organizational data favors the extraction of the evidence an organization needs for their business process improvement. In previous work, we dealt with integrating both kinds of data into a unified view that could be useful for evaluating and improving business process. In this paper, we exploit the capabilities of a data warehouse (DW) to analyze such data. We describe the DW and the process for populating their dimensions. We also show an application example from a real business process, highlighting the benefits and limitations of our approach.

Keywords: Business process improvement · Integrated process and organizational data · Data warehouse

1   Introduction

Organizations agree that their business processes (BPs) [5] are the basis of their business. However, most BPs are handled manually or are implicit in their information systems, and their data is coupled with organizational data. Although supported by a Business Process Management Systems (BPMS), the link between process data and its associated organizational data is not easy to discover.

In [3,4], we define an integrated framework for organizational data science, providing an organization with the full evidence they need to improve their operation. A key aspect of such a framework is integrating process and organizational data into a unified view to get a complete understanding of BPs execution.

In this paper, we describe an approach for taking advantage of such an integrated view with a generic and simple solution. We define a Data Warehouse (DW) capable of analyzing integrated data from many perspectives. We analyze

Supported by project "Minería de procesos y datos para la mejora de procesos en las organizaciones" funded by Comisión Sectorial de Investigación Científica (CSIC), Universidad de la República (UdelaR), Uruguay.

© Springer Nature Switzerland AG 2021
M. Golfarelli et al. (Eds.): DaWaK 2021, LNCS 12925, pp. 207–215, 2021.
https://doi.org/10.1007/978-3-030-86534-4_19

its capabilities by exploring the "Students Mobility" BP from [3], a real process from our university.

The rest of this paper is organized as follows. In Sect. 2 we describe a motivating example and, in Sect. 3, we describe our previous work. In Sect. 4, we present the DW and, in Sect. 5, we present an application example. In Sect. 6 we discuss related work. Finally, in Sect. 7 we conclude.

2   Running Example

In [3] we introduced the "Students Mobility" BP (Fig. 1), a real process from our university describing the evaluation of mobility programs for students.

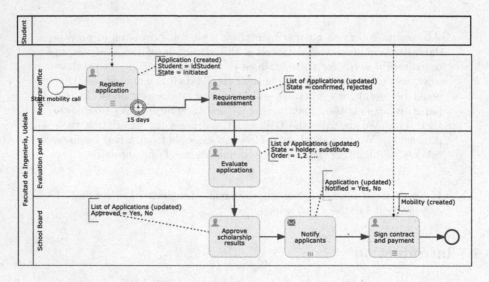

Fig. 1. Students mobility BP and its corresponding database changes

It is supported by an organizational data model containing process-specific tables: mobility program (`Program`), student's applications (`Application`) and its state (`State`), scholarships assigned (`Mobility`) and applications' courses validation (`Validation`). It also contains general data: students (`Student`) and their careers (`Career`), career courses (`Course`), the institute to which they belong (`Institute`), and teachers that work in them (`Teachers`).

The process starts when a mobility program call opens. The register office receives students' applications for 15 days. A new application is registered in the database for each one, also linking it with the corresponding open program and student. Then, the applications are checked, and rejected applications are no longer considered. Next, an evaluation board evaluates the applicants' list, marking them as holders or substitutes, and ranks them. After that, the school board assesses the resulting list, approves it, and notifies the applicants who must sign their contracts for receiving their payment and start the mobility.

3   Integrating Process Data and Organizational Data

In [3] we dealt with integrating process data and organizational data, as part of our integrated framework [4], based on three perspectives. We also discussed issues regarding the ETL process and the matching algorithm we proposed. The **raw data perspective** involves BPMS managing processes and an organizational database with its corresponding change log relating its data with BPMS events through time. The **model perspective** involves defining an integrated metamodel, whose schema is depicted in Fig. 2. It is composed of a process and a data view, both with two levels of information: the definition of elements and their instances. The *process-definition quadrant* describes processes, their corresponding elements (e.g., tasks), user roles, and variables defined in design time. The *process-instance quadrant* represents a process execution, i.e., process cases which have element instances, some of them performed by users, that manipulate specific instances of variables. The *data-definition quadrant* describes a generic data model with entities and their corresponding attributes. Finally, the *data-instance quadrant* describes instances of these elements containing specific values that evolve through time. The metamodel is populated using data extracted from the raw data perspective. Once a model is loaded, a matching mechanism deals with discovering the existing connections between process data and organizational data. The **DW perspective** structures the information in the metamodel into a DW, providing means for exploiting such integrated data.

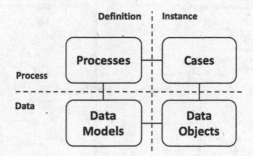

Fig. 2. High-level definition of the metamodel (from [3])

4   Data Warehouse Approach

The DW was built to provide the basis for analyzing the integrated process data and organizational data. A key element is providing a generic approach to its design. Thus, no domain-specific elements regarding the process data or organization are used; It is based solely on the relationships between process data and organizational data previously discovered in the metamodel. The DW logical design corresponds to a star schema whose dimensions and measures are described next. The main dimensions are based on the metamodel quadrants.

Process Dimensions. The process-definition dimension (Fig. 3(a)) takes concepts from the quadrant with the same name. The elements represent a process definition with its elements and variable definitions. In the hierarchy, variable definitions are in the lower level, and roll-up is allowed to element definitions and up to the process definition level. The process-instance dimension (Fig. 3(b)) considers case identification, element instances, and variable instances. Variable instances are the lower level, and roll-up is allowed to element instances and up to the process instances level.

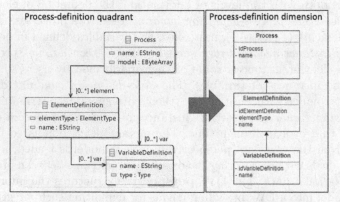

(a) Process-definition quadrants and dimensions

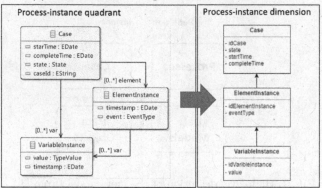

(b) Process-instances quadrants and dimensions

Fig. 3. Process quadrants and their corresponding DW dimensions

Data Dimensions. Data dimensions are also based on the corresponding meta-model quadrants. The data-definition dimension (Fig. 4(a)) considers entity definition and its corresponding attributes. The lower level is represented by attributes from which roll-up is allowed to entities in the defined hierarchy. The data-instance dimension (Fig. 4(b)) considers entity instances and corresponding attribute instances. Attribute instances represent the lower level in the defined hierarchy from which roll-up is allowed to related entity instances.

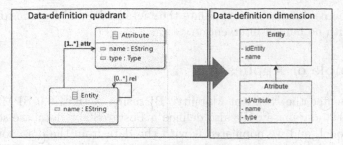

(a) Data-definition quadrants and dimensions

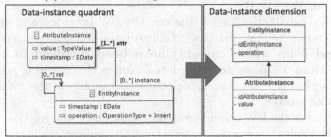

(b) Data-instances quadrants and dimensions

Fig. 4. Data quadrants and their corresponding DW dimensions

Other Dimensions. We defined two cross-quadrant dimensions. The User dimension takes concepts from the process quadrants regarding roles and user for process execution. The lowest level corresponds to users from which roll-up is allowed to role definition. The EntityRelation dimension takes concepts from the data quadrants regarding relationships between entities, i.e., foreign keys. The lower level is represented by relations between entity instances from which roll-up is allowed to corresponding relations between entity definitions. Finally, we defined a Time dimension to carry out analysis based on time ranks.

Measures. The fact table relates the dimensions mentioned before. We include in the fact table process duration and element duration to analyze execution times for both process and elements, i.e., cases execution from start to finish and elements execution within all cases. We also included the value of attributes.

Discussion. Crossing data from the defined dimensions, we can answer questions from every critical perspective: processes performed, data managed, people involved, and time, relating these perspectives for the analysis. The analysis we provide is also generic in terms of domain objects. To perform a specific domain analysis, we need to include dimensions regarding specific concepts and relationships between organizational data. Although it would be interesting and would enhance the domain analysis, keeping the DW generic provides several advantages. In particular, it focuses on analyzing the relationship between data dimensions. Also, we can define a mostly automated ETL process to obtain data from the source databases to populate the metamodel and from it to populate the DW, with no human intervention.

5   Example of Application

We implemented the "Student Mobility" BP using Activiti 6.0 BPMS[1] with a MySQL 8.0.20 database[2]. We also defined a PostgreSQL[3] database schema for the metamodel, and we populated it with the data from the data sources. We use Pentaho[4] for its construction and the implementation of the ETL process required for building the DW.

To show the capabilities for BP analysis the DW provides, and due to space reasons, we chosen to describe how to answer the following question: **which courses (and from which careers) have been involved in cases that took more than 15 days to complete?** Notice that there is no individual data source within the raw data perspective containing all the information to provide such an answer.

The query over the DW should cross data filtered by the relation validation-course, which defines the courses included in the applications over the process, along with the case id and the corresponding attributes. Then, the filter for the query includes the "existRelation" attribute from the "Entityrelation" dimension set to "true". The "Entityrelation" dimension is also used as a column filtered by the "Name" attribute, selecting the "validation-course" relation over organizational data. As rows, the selection includes the corresponding "id" attribute from the "Entityrelation" dimension, the "idcase" attribute from the "ProcessInstance" dimension, the "Entity name" and "Attribute name" attributes from the "DataDefinition" dimension (filtered by attributes "idcourse" and "idcareer"), and attribute value from the "DataInstance" dimension. On the "Measures dimension", the "Process duration" measure is selected and filtered by duration over 200.000 ms (in this example, 15 days are equivalent to 200.000 ms).

Figure 5 shows a screenshot of Pentaho providing the results for the example question. It can be seen that cases are related to validation and the identification of the courses included in the corresponding student's application. As the career is not directly associated with the case, it is recovered from the relation course-career existing only in the organizational database.

6   Related Work

In [3] we discussed proposals related to the integration of process data and organizational data. Most of them observe the problem from the process mining

---

[1] https://www.activiti.org/.

[2] https://www.mysql.com/products/community/.

[3] https://www.postgresql.org/.

[4] https://www.hitachivantara.com/en-us/products/data-management-analytics/ pentaho-platform.html.

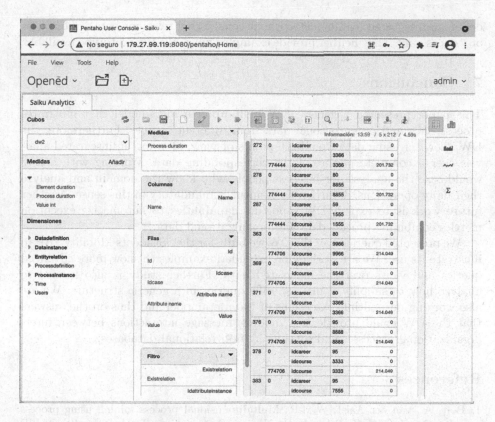

Fig. 5. DW result answering the example question

perspective, not exploiting both sources of information altogether, e.g., business-level activities are omitted, or the analysis is focused on process events.

In [9], the authors define a framework for an in-depth exploration of process behavior combining information from the event log, the database, and the transaction log. Although they analyze how elements can be connected, they only define how this information can be exploited from a theoretical perspective. In [7], the authors perform a literature review to identify the connection between process mining and DW. Most works are focused on using DW from the perspective of process mining, as the one in [1]. Few works consider both sources of information. In [8] the authors propose a method for constructing a DW based on the information used within the BP. This approach is discussed in some works [2,6] since it allows a fine-grained analysis of domain data, but at the same time, it requires the definition of a new DW and their reporting solution for each BP. In [2], the authors define a generic DW process model and a strategy for mapping business data (kept outside the DW) and the DW process model. This proposal needs a deep understanding of the business logic to match the business data. Moreover, the proposal is proprietary, and no technological solution is available. Finally, in [6], the authors also provide a generic solution quite close to our proposal. They also define an integrated metamodel, without considering relations between business objects as ours. They neither provide examples nor make the prototype available.

7   Conclusions

This paper presented a generic DW built upon a metamodel that integrates process data and organizational data from different sources. Furthermore, the DW allows crossing data from both points of view, allowing an integrated analysis over a complete picture of the corresponding data. Working with a DW would ease the analysis for business people who know the domain and analyze their daily operations. A further assessment is mandatory in this sense. Another future work is to exploit DW reporting capabilities to automatize execution metrics combining process data and organizational data.

We presented queries over the DW we built for the "Students Mobility" BP to illustrate its use. We are working on extended examples to show more advanced queries and views over the integrated data. Further study is also needed to analyze how to simplify domain-specific queries in a generic structure. We are also working on multiple extensions of the metamodel and thus on the answers that the DW could provide, e.g., adding message interactions between inter-organizational processes and considering non-relational databases.

References

1. Bolt, A., van der Aalst, W.M.P.: Multidimensional process mining using process cubes. In: Gaaloul, K., Schmidt, R., Nurcan, S., Guerreiro, S., Ma, Q. (eds.) CAISE 2015. LNBIP, vol. 214, pp. 102–116. Springer, Cham (2015). https://doi.org/10.1007/978-3-319-19237-6_7
2. Casati, F., Castellanos, M., Dayal, U., Salazar, N.: A generic solution for warehousing business process data. In: 33rd International Conference on Very Large Data Bases (VLDB), pp. 1128–1137. ACM (2007)
3. Delgado, A., Calegari, D.: Towards a unified vision of business process and organizational data. In: XLVI Latin American Computing Conference (CLEI), pp. 108–117. IEEE (2020)
4. Delgado, A., Marotta, A., González, L., Tansini, L., Calegari, D.: Towards a data science framework integrating process and data mining for organizational improvement. In: 15th International Conference on Software Technologies (ICSOFT), pp. 492–500. ScitePress (2020)
5. Dumas, M., La Rosa, M., Mendling, J., Reijers, H.A.: Fundamentals of business process management. In: Lecture Notes in Business Information Processing, Springer, Heidelberg (2018). https://doi.org/10.1007/978-3-662-56509-4_9
6. Kassem, G., Turowski, K.: Matching of business data in a generic business process warehousing. In: 2018 International Conference on Computational Science and Computational Intelligence (CSCI), pp. 284–289 (2018)
7. Križanić, S., Rabuzin, K.: Process mining and data warehousing - a literature review. In: 31th Central European Conference on Information and Intelligent Systems (CECIIS), pp. 33–39. University of Zagreb, Faculty of Org. and Informatics Varaždin (2020)

8. Sturm, A.: Supporting business process analysis via data warehousing. J. Softw. Evol. Process **24**(3), 303–319 (2012)
9. Tsoury, A., Soffer, P., Reinhartz-Berger, I.: A conceptual framework for supporting deep exploration of business process behavior. In: Trujillo, J.C., et al. (eds.) ER 2018. LNCS, vol. 11157, pp. 58–71. Springer, Cham (2018). https://doi.org/10.1007/978-3-030-00847-5_6

Smart-Views: Decentralized OLAP View Management Using Blockchains

Kostas Messanakis, Petros Demetrakopoulos, and Yannis Kotidis[✉]

Athens University of Economics and Business, Athens, Greece
kotidis@aueb.gr

Abstract. In this work we explore the use of a blockchain as an immutable ledger for publishing fact records in a decentralized data warehouse. We also exploit the ledger for storing smart views, i.e. definitions of frequent data cube computations in the form of smart contracts. Our techniques model and take into consideration existing interdependencies between the smart views to expedite their computation and maintenance.

1 Introduction

Many blockchain infrastructures utilize some form of "smart contracts", as pieces of code that are embedded in the ledger implementing binding agreements between parties that can, for instance, auto-execute when certain conditions are met. Building on smart contracts, in this work we envision a decentralized data warehouse implemented on top of a blockchain, where smart contracts are used as the means to describe data processing over the collected multidimensional data. In essence, these contracts define aggregate views over the data and we refer to them as smart views. Smart views can trigger computations enabling complex data analysis workflows.

Using the smart views paradigm, organizations can implement a decentralized data warehouse using the blockchain ledger for publishing its raw data (fact records) as is depicted in Fig. 1. Smart views feed from this data and implement, on local nodes, smaller data marts [7] that are used for specific analytical purposes. For example, local node 1 in the figure can be utilized by data scientists working on sales data, while local node n by executives analyzing inventory data. Both user groups will need access to the distributed data warehouse repository. Each user group may have its own analytical queries focusing on specific aspects of the data. Their needs will be accommodated by their local node that (i) syncs data updates from the shared data warehouse ledger, (ii) manages local user inquires via the smart views API we provide and (iii) caches local results so that subsequent queries are expedited. Queries that are common among the groups will be shared at a semantic level, as their descriptions will be also stored in the distributed ledger.

Smart views, much like traditional views in a database system, are virtual up to the moment when they are instantiated by a user request. On each node, we utilize a data processing engine (that can be a relational database or a big data

M. Golfarelli et al. (Eds.): DaWaK 2021, LNCS 12925, pp. 216–221, 2021.
https://doi.org/10.1007/978-3-030-86534-4_20

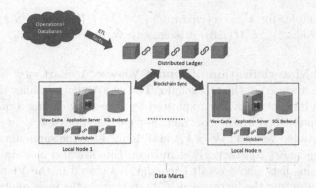

Fig. 1. Using Smart Views for building a decentralized Data Warehouse.

platform) for doing the heavy lifting of computing their content. Additionally, an in-memory database acts as a View Cache, providing fast access to their derived data. This cache is used to serve multiple user requests on a node, orders of magnitude faster than what a blockchain could provide. In addition to engineering a solution that scales blockchains for processing data warehouse workloads, we also propose and compare algorithms that can manage effectively the smart views content cached in the in-memory database.

2 Smart Views

Preliminaries and Definitions: In order to simplify the notation used, we assume that the data warehouse implements a set of dimensions $D = \{d_1, d_2, \ldots d_n\}$ and a single measure m. A smart view V is defined by projecting the fact table records on a subset $D_V \subseteq D$ of the available dimensions and computing a function F_V on their measure values. For this discussion, we assume that F_V is one of the commonly used distributive or algebraic functions such as $MAX()$, $MIN()$, $SUM()$, $COUNT()$, $AVG()$, $STDEV()$, $top_k()$, etc. Optionally, the view may contain a set of conjunctive predicates P_V on the dimensions in set $D - D_V$ that can be used, for instance, in order to perform a *slice* operation [5]. Thus, a smart view can be described as a triplet $V = \{D_V, F_V, P_V\}$.

As an example, assume a data warehouse schema with three dimensions $D = \{customer, product, store\}$ and one measure $amount$. A smart view may compute the total sales per customer and product for a specific store S. Then, $D_V = \{customer, product\}$, $F_V = SUM()$ and $P_V = \{'store = S'\}$.

Given two views $V_i = \{D_{V_i}, F_{V_i}, P_{V_i}\}$ and $V_j = \{D_{V_j}, F_{V_j}, P_{V_j}\}$, we say that view V_i is more detailed than view V_j ($V_j \preceq V_i$) iff the following conditions hold

1. All dimensions in view V_j are also present in view V_i, i.e. $D_{V_j} \subset D_{V_i}$.
2. F_i and F_j are the same aggregation function.
3. Database query $Q_j(P_{V_j})$ is contained in query $Q_i(P_{V_i})$, for every state of the data warehouse fact table DW. Query $Q_i(P_{V_i})$ is expressed in relational algebra as: $\pi_{D_{V_i}}(\sigma_{P_{V_i}}(DW))$, and similarly for $Q_j(P_{V_j})$.

As an example, for $V_1 = (\{customer\}, SUM(), \{'store = S'\})$ and $V_2 = (\{customer, product\}, SUM(), \emptyset)$, it is easy to verify that $V_1 \preceq V_2$.

On-Demand Materialization of Smart Views: A smart view is defined via a smart contract stored in the blockchain. The smart view remains virtual, up to the time when its content is first requested by a user, by calling a *materialize()* API function published by the contract. For each smart view V, we maintain a list of more detailed views V_i ($V \preceq V_i$) also defined for the same data warehouse schema via the smart contract. By definition, this list also contains view V. A view V_i from this list, whose result $cached(V_i)$ is stored in the View Cache can be used for computing the instance of V, by rewriting the view query to use this result, instead of the fact table DW. Nevertheless, we should also provision for fact table records that have been appended to the blockchain after the computation of the cached instance of V_i. Let t_i be the timestamp of the computation of the cached result of V_i. Let $deltas(t_i)$ denote the records that have appeared in the blockchain after timestamp t_i. After retrieving the cached, stale copy of view V_i and the deltas, the contents of the view are computed by executing the view query over the union of the data in $cached(V_i)$ and $deltas(t_i)$. This functionality is provided by the SQL backend. The result of the smart view is returned to the caller of function *materialize()* and is also sent to the View Cache.

We estimate the cost at the backend as $w_{sql} \times (size_{cached(i)} + size_{deltas(i)})$, for some weight parameter w_{sql} that reflects the speed of the SQL backend. Retrieval of records in $deltas(t_i)$ is performed by the application server that is also running a blockchain node. We estimate this cost as $w_{ledger} \times size_{deltas(i)}$.

In total, the cost of using smart view V_i in order to materialize view V is estimated as

$$cost(V_i, V) = w_{sql} \times (size_{cached(i)} + size_{deltas(i)}) + w_{ledger} \times size_{deltas(i)}$$

which can be simplified as ($w_{ledger} >> w_{sql}$)

$$cost(V_i, V) = \alpha \times size_{deltas(i)} + size_{cached(i)}$$

for some constant $\alpha >> 1$. The formula suggests that stale results in the cache are less likely to be used for future materializations of smart views.

Cost-Based View Eviction: Let $\mathcal{V}$ denote the set of views that have been requested earlier by the application. For each view $V \in \mathcal{V}$ we also maintain the frequency f_V of calling function *materialize(V)*. For the calculations that follow, we remove from set $\mathcal{V}$ those smart views that can not be computed using results in the View Cache. Let $costMat(V)$ denote the minimum cost of materializing view V and $\mathcal{VC}$ denote the set of all views presently stored in the View Cache. Then,

$$costMat(V, \mathcal{VC}) = \min(\min_{V_i \in \mathcal{VC}: V \preceq V_i} cost(V_i, V), costBC)$$

where $costBC$ is the cost of computing V directly from the blockchain data.

The *amortized displacement cost* $dispCost(V_i)$ of cached result V_i is defined as the cumulative increase in the materialization cost of all smart views $V \preceq V_i$, in case view V_i is evicted, divided by its size. Thus,

$$dispCost(V_i) = \frac{\sum_{V \in \mathcal{V}: V \preceq V_i} f_V \times (costMat(V, \mathcal{VC}) - costMat(V, \mathcal{VC} - V_i))}{size_{cached(i)}}$$

Our proposed cost-based view eviction policy (COST) lists the views in the cache in increasing order of their amortized displacement cost. The eviction process then discards views from the cache by scanning this list until enough space is generated for storing the new result.

Eviction Based on Cube Distance. In order to capture locality exhibited between successive requests for smart views, we propose an alternative eviction policy (DIST) based on the data cube distance metric. The Data Cube lattice [9] contains an edge (V_i, V_j) iff $V_i \preceq V_j$ and $|D(V_j)| = |D(V_i)| + 1$. The Data Cube distance ($distCube$) in the lattice between two views is defined as the length of the shortest path between the views. As an example, $distCube(V_{product,store}, V_{product,customer}) = 2$. For a newly computed view V, we order the views in the cache in decreasing order based on their distance from V and discard views until enough space is generated as a result of these evictions.

3   Experiments

In this section we provide experimental results from a proof-of-concept implementation of our proposed system. Ganache was used for setting up a personal Ethereum Blockchain and implementing the Smart Views API using Solidity contracts. We used MySQL as a relational backend and Redis for implementing and View Cache. The application server functionality was written in Node JS.

We used synthetically generated data using 5 dimensions with cardinality 1,000 integer values each. Each fact table record contained five keys (one for each dimension) and one randomly generated real value for the measure. In order to evaluate the performance of the View Cache we generated three workloads consisting of 1000 queries (views) each. In the first, denoted as W_{olap}, we started from a random view in the lattice and each subsequent query was either a roll-up or drill-down from the previous one. The second workload W_{random} contained randomly selected views from the lattice (with no temporal locality). For the last workload, W_{mix}, a subsequent query was a roll-up or a drill-down from the previous query or was a new randomly generated one. In order to test the View Cache we implemented three eviction policies. The first two, cost based (denoted as COST in the graphs), and distance-based (denoted as DIST in the graphs) are discussed in Subsect. 2. We also implemented Least Recently Modified (LRM), a simple policy that selects as a victim the cached view that has not been modified for the longest time in the cache.

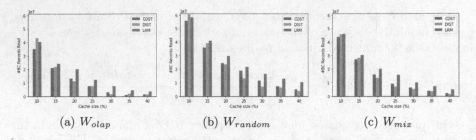

(a) W_{olap} (b) W_{random} (c) W_{mix}

Fig. 2. Number of records fetched from the blockchain.

For each dataset, we started with a blockchain containing 100000 records. We then run the experiment for 1000 epochs. During each epoch we inserted 100 new records and executed a query from the respective workload. Thus, at end of the run, the blockchain stored 200000 fact table records. In Fig. 2 we plot the number of blockchain records read for the three different eviction policies and the three workloads used, as we varied the View Cache size from 10% up to 40% of each workload result size in increments of 5%. As expected, the W_{olap} workload resulted is much fewer reads from the blockchain, as successive queries target nearby views in the Data Cube lattice due to the roll-up and drill-down operations. On the other hand, W_{random} contains queries with no temporal locality, resulting in fewer hits in the cache. Nevertheless, the benefits of the cache are significant, even for this random workload. Compared to not using the cache, even the smaller cache size (10%) with the COST policy reduces the total number of records fetched from the blockchain from 150M down to 55M, i.e. a reduction of 63%. The largest cache tested (40%) reduced this number by 97%. The results for workload W_{mix} are in between, as expected. Comparing the three eviction policies, we observe that the COST policy works better in constrained settings (smaller cache sizes), as it avoids flashing from the cache, smart view results that help materialize many other views in the workload. The DIST policy is often better for intermediate cache sizes, while LRM performs worst, as it does not take into consideration the interdependencies between the cached view results.

4   Related Work

Existing studies [2] show that blockchain systems are not well suited for large scale data processing workloads. This has been our motivation for decoupling in the proposed architecture the definition and properties of the smart views (stored in the ledger) from their processing and maintenance that are handled by appropriate data management modules. Our architecture differs from existing solutions that either add a database layer with querying capabilities on top of a blockchain [1,3] or, extend databases and, in some cases NoSQL, systems with blockchain support [6,12,13]. A nice overview of these systems is provided in [11]. There are also proposals that aim to build verifiable database schemas [4,14] that can be shared among mutually untrusted parties.

The work on smart views is motivated by our previous work on view selection in data warehouses [8–10]. Unlike traditional data warehouses, the raw data required for deploying the smart views is stored in the ledger, while processing and management of their derived aggregate results happens off-the-chain. As a result, in this work we proposed a new cost model, tailored for our modular architecture and implemented new view cache policies based on this model.

5   Conclusions

In this work we proposed a decentralized data warehouse framework for defining, (re)using and maintaining materialized aggregate smart views using existing blockchain technology. Our experimental results, based on a proof-of-concept prototype we built demonstrate the effectiveness of smart views in reducing significantly the overhead of fetching stored data from the distributed ledger.

References

1. AWS. Amazon Quantum Ledger Database (QLDB)
2. Dinh, T.T.A., Wang, J., Chen, G., Liu, R., Ooi, B.C., Tan, K.: BLOCKBENCH: a framework for analyzing private blockchains. In: SIGMOD. ACM (2017)
3. El-Hindi, M., Binnig, C., Arasu, A., Kossmann, D., Ramamurthy, R.: Blockchaindb - A shared database on blockchains. In: Proceedings of the VLDB Endowment (VLDB) (2019)
4. Gehrke, J., et al.: Veritas: shared verifiable databases and tables in the cloud. In: CIDR, Asilomar, January 2019
5. Gray, J., Bosworth, A., Layman, A., Pirahesh. H.: Data cube: a relational aggregation operator generalizing group-By, cross-tab, and sub-total. In: ICDE, pp. 152–159 (1996)
6. Helmer, S., Roggia, M., Ioini, N.E., Pahl, C.: Ethernitydb - integrating database functionality into a blockchain. In New Trends in Databases and Information Systems - ADBIS Short Papers and Workshops (2018)
7. Inmon, W.H.: Building the Data Warehouse. QED Information Sciences Inc., Wellesley (1992)
8. Kotidis, Y., Roussopoulos, N.: An alternative storage organization for ROLAP aggregate views based on Cubetrees. In: Proceedings of SIGMOD (1998)
9. Kotidis, Y., Roussopoulos, N.: DynaMat: a dynamic view management system for data warehouses. In: Proceedings of SIGMOD, pp. 371–382 (1999)
10. Kotidis, Y., Roussopoulos, N.: A case for dynamic view management. ACM Trans. Database Syst. 26(4), 388–423 (2001)
11. Raikwar, N., Gligoroski, D., Velinov, G.: Trends in development of databases and blockchain. In: SDS, pp. 177–182. IEEE (2020)
12. Sahoo, M.S., Baruah, P.K.: Hbasechaindb - a scalable blockchain framework on hadoop ecosystem. In: Supercomputing Frontiers, Singapore, March 26 20 (2010)
13. Schuhknecht, F.M., Sharma, A., Dittrich, J., Agrawal, D.: chainifydb: how to get rid of your blockchain and use your DBMS instead. In: CIDR (2021)
14. Zhang, M., Xie, Z., Yue, C., Zhong, Z.: Spitz: a verifiable database system. Proc. VLDB Endow. 13(12), 3449–3460 (2020)

A Workload-Aware Change Data Capture Framework for Data Warehousing

Weiping Qu[1], Xiufeng Liu[2(✉)], and Stefan Dessloch[1]

[1] Heterogeneous Information Systems Group, University of Kaiserslautern,
Kaiserslautern, Germany
{qu,dessloch}@informatik.uni-kl.de
[2] Department of Technology, Management and Economics,
Technical University of Denmark, 2800 Lyngby, Denmark
xiuli@dtu.dk

Abstract. Today's data warehousing requires continuous or on-demand data integration through a Change-Data-Capture (CDC) process to extract data deltas from Online Transaction Processing Systems. This paper proposes a workload-aware CDC framework for on-demand data warehousing. This framework adopts three CDC strategies, namely trigger-based, timestamp-based and log-based, which allows capturing data deltas by taking into account the workloads of source systems. This paper evaluates the framework comprehensively, and the results demonstrate its effectiveness in terms of quality of service, including throughput, latency and staleness.

Keywords: Change-Data-Capture · Workload-aware · Data
warehousing · Quality of service

1 Introduction

In data warehousing, the ETL (Extract-Transform-Load) process cleans, processes and integrates data into a central data warehouse via various data processing operators [4]. The CDC is a key component of an ETL process to achieve continuous integration of incremental data. It can synchronise updates from a source system, for example committed transactions from an online transaction processing system (OLTP), to a data warehouse [2]. However, CDC processes require some level of initial integration cost at runtime, which may compete for resources with the source system, with negative implications. Depending on the implementation of CDC processes (e.g. based on logs, triggers and timestamps) and the configuration of the processes (e.g. number of concurrent CDC threads and execution frequencies), the impact on performance can be quite significant in terms of quality of service (QoS), such as transaction throughput. Therefore, OLTP and CDC workloads should be reasonably scheduled to meet QoS requirements, for example, based on resource utilisation. In the test in this paper, we

© Springer Nature Switzerland AG 2021
M. Golfarelli et al. (Eds.): DaWaK 2021, LNCS 12925, pp. 222–231, 2021.
https://doi.org/10.1007/978-3-030-86534-4_21

refer to QoS for OLTP workloads as transaction quality (QoT), and QoS for OLAP queries as analysis quality (QoA).

   In recent years, some attempts have been devoted to load-based, near-real time or on demand data warehousing. For example, Thiele et al. developed a partition-based workload scheduling approach for a live data warehouse [6]. Issam et al. proposed a feedback-controlled scheduling architecture for a real-time data warehousing [1]. Shi et al. proposed a QoS-based load scheduling algorithm for query tasks [5]. Vassiliadis et al. introduced the concept of near real-time ETL for online data warehousing, which is capable of generating deltas to update the target data warehouse incrementally [8]. Thomsen et al. proposed a righ-time ETL that supports data request to source systems by specifying data freshness in a data warehouse [7], and Liu et al. extended this work to support update, insert and delete operations and state management [3]. However, these efforts have built individual tools or platforms for data warehousing for a specific purpose, e.g. with a shorter loading time, but without sufficiently addressing the requirements of QoD and QoA. It is therefore necessary to build an "intelligent" framework to extract deltas according to various workloads of source systems.

   To address this need, in this paper we propose a workload-aware CDC framework, which integrates three classical CDC methods, including trigger-based, log-based and timestamp-based methods for data extraction. With this framework, users can tune their CDC methods to meet different freshness demands with given delays at runtime. We design the runtime configurations, by which users can "gear" the CDC processes at runtime according to the the QoA requirement, i.e., increase/decrease scheduling frequencies or execution parallelism.

2   Problem Analysis

The end-to-end latency of delta synchronisation starts with capturing source deltas, their buffering in the staging area (CDC delay) and ends with merging the deltas with the target data warehouse tables. This paper focuses only on the CDC delay, ignoring the transformation and loading phases, as they do not influence the QoT.

   A source-side transaction T_i successfully commits at time $t_c(T_i)$ in an OLTP source system while it is extracted and recorded at time $t_r(T_i)$ in the staging area. The freshness of the OLTP source at time t is denoted as $t_{f_s}(t)$ which is the latest commit time, i.e., $\max(t_c(T_i))$. The freshness of data staging area at time t is denoted as $t_{f_d}(t)$ which is the newest commit time of the updates captured in the staging area, i.e., $\max(t_r(T_i))$. A CDC request C_i arrives at time $t_a(C_i)$ with a time window $t_w(C_i)$. This time window is the maximum time period which the C_i can tolerate to wait for the CDC to deliver the complete set of deltas, thus avoiding infinite suspension of OLAP queries due to limited resources in OLTP systems. The deadline $d(C_i)$ for the request C_i is calculated as the sum of the request arrival time $t_a(C_i)$ and the time window of the request $t_w(C_i)$. In addition to the deadline specified in the request C_i, users can explicitly inform the CDC the desired accuracy $t_{f_s}(C_i)$ for an analytical query by invoking ensureAccuracy(...) through JDBC connection [7]. To fulfil the expected freshness, the CDC processes should guarantee the following conditions, i.e., when the maximal waiting period is due, the freshness of the staging area should be equal to or greater than the specified freshness demand in C_i. Otherwise, the CDC request expires and returns the results with specific staleness satisfying the following conditions, $t_{f_d}(t) \geq t_{f_s}(C_i)$ and $t \leq d(C_i)$.

The staleness of request C_i is $S(C_i)$, the difference between the expected freshness and the actual freshness returned by C_i. If C_i is successfully fulfilled, the difference is zero, i.e., $S(C_i) = t_{f_s}(C_i) - t_{f_d}(t)$ and $t = d(C_i)$ if C_i expires, otherwise $S(C_i) = 0$. The latency $L(C_i)$ is the time taken by the CDC to answer the request C_i. If the request expires, $L(C_i)$ is equal to the time window of C_i, i.e., $L(C_i) = t_w(C_i)$ if C_i expires, otherwise $L(C_i) = t_r(T_i) - t_a(C_i)$.

To better illustrate the impact of CDC processes on original OLTP systems, we conducted an experiment which compared the QoT metric τ with the QoA metric S (average staleness) using trigger-based CDC with two configuration settings: *high gear* (schedule interval 20 ms, 2 threads) vs. *low gear* (schedule interval 640 ms, 1 thread). As shown in Fig. 1, although the average staleness S in the high gear performs better than that in the lower gear, the transaction throughput τ is way worse and not satisfactory for OLTP workloads, i.e., delivering fresher deltas incurs significant overhead on transaction executions. This experiment demonstrates the trade-off between QoT and QoA and the necessity of a workload-aware CDC for trade-off balancing.

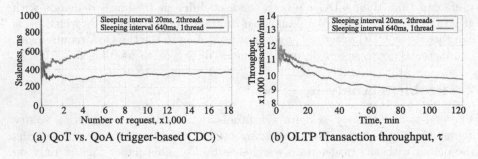

(a) QoT vs. QoA (trigger-based CDC)   (b) OLTP Transaction throughput, τ

Fig. 1. The study of an illustrative CDC process example

To quantify the effectiveness of our workload-aware CDC, we analyse the QoT and the QoA together and consider their trade-off, defined as bellow:

$$Gain(G) = w \times \alpha \times \tau + (1 - w) \times \beta \times \frac{1}{S \times L \times Er} \tag{1}$$

Since τ drops with shrinking staleness S, latency L or expiration ratio Er, the QoT value is directly set as τ while the QoA value is calculated as the reciprocal of the production of S, L and Er. To compare QoT and QoA values, we introduce two normalization coordinates α and β for them. User can set the weights for the qualities of services, i.e., w, which can be adjusted at runtime. The performance gain value G is the sum of weighted, normalized values of both QoT and QoA. The objective of workload-aware CDC is to approximate $\max(G)$ at any time.

3   Proposed Workload-Aware CDC Framework

Figure 2 shows the overall architecture. On the left-hand side, there are three CDC approaches: trigger-based, timestamp-based and log-based. The source deltas are extracted by the workload-aware CDC components (middle of the figure) and propagated to an staging area via a messaging broker, e.g., Kafka (right). Each CDC component runs as a separate ETL process, consisting of a thread agent, a job queue and a thread pool worker. The agent thread generates CDC tasks for a specific delta at a rate that can be customised at runtime (step 1). Each task is immediately added to the task queue (step 2), then executed to fetch (step 3) and produce (step 4) source deltas when the thread pool has an idle thread. To achieve workload awareness, the traditional Data Warehouse (DWH) manager is extended with additional workload-aware CDC logic. The DWH manager keeps calculating the performance gains for the current QoT and QoA. The QoT, i.e., transaction throughput τ, is continuously reported to the DWH manager by a source-side *monitor* while the QoA metrics, i.e., average staleness S, latency L and expiration ratio Er, are measured over the final status of the CDC requests sent to the DWH manager. With varying freshness demands and deadlines set in incoming CDC requests, the DWH manager triggers the CDC executions to switch to new gears immediately, e.g., by raising/shrinking frequency or resource pool size, when the performance gain drops.

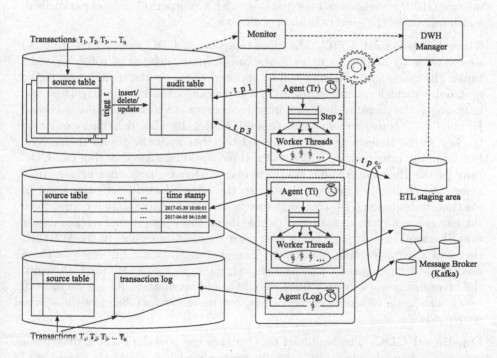

Fig. 2. System architecture of the proposed workload-aware CDC framework

The following describe the specific CDC methods for workload-aware:

Trigger-Based CDC. The trigger-based CDC is shown at the upper-left corner of Fig. 2. Triggers are the functions which are defined manually to specify how the updates are tracked from committed transactions over source tables and fired automatically whenever certain events occur, such as after/before insert, delete or update operation. The captured updates are recorded in a separate table, called *audit* table, which becomes visible when the original transactions commit. The trigger-based CDC component (at the upper-middle of Fig. 2) fetches deltas directly from the audit table by issuing SQL queries. To identify new deltas, the audit table schema is designed to track the sequence ids of transactions. The agent thread periodically reads the latest id and compares it with the cached maximum sequence id read from the last time. If a new sequence id is found, a task is constructed with a sequence id range of $(oldid, newid)$, i.e., lower and upper bounds of the sequence ids of those new delta tuples that need to be fetched. When being processed by a worker thread, a delta-fetching SQL query is generated with this range as a predicate condition in the WHERE clause. The gears of the trigger-based CDC are characterized by two properties: *scheduling period* i.e., how long the agent stays idle in a period, and *worker parallelism*, i.e., the size of the worker thread pool. High gear is set with short scheduling period and high parallelism to keep the deltas up-to-date in the staging area. However, it imposes more SQL-query load on the audit table and contends for the resources and the locks with original OLTP transactions. Low gear is set with a long period and low parallelism, which influences the sources in an opposite way.

Timestamp-Based CDC. The timestamp-based CDC extends the original source table by adding an extra *timestamp* column, instead of using an audit table. The timestamp of a tuple will be updated whenever the tuple is inserted or updated, by which the modifications of a tuple can be tracked directly. However, indexing on the timestamp column introduces extra index-maintenance overhead for the OLTP transactions similar to audit tables. Besides, deletions cannot be tracked as the timestamp will be deleted together with the tuples themselves. Furthermore, intermediate updates of a tuple could be missed during two CDC runs as the timestamp value could be overwritten by new ones at any time before the change is captured. Similar to the trigger-based CDC, the agent of the timestamp-based CDC component caches the latest timestamp recorded from the last run and compares it with the maximum timestamp found in the current run. The sequence id range is replaced with a timestamp range in the WHERE clause of a delta-fetching SQL query. The gear settings are the same as those used in the trigger-based CDC. In addition, the more frequently the CDC runs, the more deltas are captured from a tuple as old versions would not be missed, which also leads to more concise deltas, but more network bandwidths at the source side.

Log-Based CDC. The log-based CDC makes use of a database management system's logging functionality (usually write-ahead logs (WAL)). Before running an INSERT/UPDATE/DELETE statement of a transaction, a log record is created and written into an in-memory WAL buffer. When the transaction commits, the buffered log records are encoded as binary stream and flushed (through *fsync*) into a disk-based WAL segment file. Log records of concurrent transactions are sequentially written into the transaction log in commit order and additional optimization work batches up log records from small transactions in one *fsync* call, which amortizes disk I/O. The log-based method reads the WAL segment files sequentially from disk (with I/O costs) and decodes the binary WAL stream through vendor-specific API into a human-readable format (using extra CPU cycles). Sequential read is fast as it saves disk seeking time and rotational delay. With a *read-ahead* feature supported by most of operating systems, subsequent blocks are pre-fetched into the page cache, thus reducing I/O. As compared to the trigger-based and timestamp-based CDCs, the main benefit of the log-based CDC is that there is no specific CDC extension needed, i.e., no audit table or timestamp column at the source side, which can eliminate the resource/lock contentions. The limitation of log-based approach is that the log entries can only be read out in an sequence order, thus prohibiting CDC worker parallelism. Therefore, gears of the log-based CDC are only determined by the scheduling period set in the agent thread, which, for example, affects source-side I/O bandwidths.

4 Experiments

The experiments are conducted on a cluster with 6 nodes, each of which is equipped with 2 Quad-Core Intel Xeon Processor E5335, 4x2.00 GHz, 8 GB RAM, 1TB SATA-II disk, Gigabit Ethernet. One node is used as the data source node with PostgreSQL. The open-source database performance testing tool BenchmarkSQL is deployed in one of the rest nodes and runs continuously. Three nodes from the rest are used to simulate the CDC processes with the same settings and competed for resources with the BenchmarkSQL processes on PostgreSQL node. The last node runs a CDC-request simulator process to issue CDC requests with varying freshness demands and deadlines, which distinguishes different priority patterns in different time slots.

Evaluation of the Trigger-Based CDC. For the trigger-based CDC, we compare the results of the following settings: three constant settings including *high gear* (sleeping interval 20 ms, 2 threads), *middle gear* (sleep interval 160 ms, 2 threads), *low gear* (sleeping interval 320 ms, 1 thread) and a *workload-aware* scheme with mixed gears.

As shown in Fig. 3, we find that neither the high gear nor the low gear is able to balance the QoS for OLTP and OLAP workloads. When the high gear is used, it can achieve the lowest latency (20 ms), keep the most up-to-date deltas in the staging area (only about 10 ms delay) and complete all CDC requests on time. However, its transaction throughput is the lowest, only about 36 K/min. The slow gear presents the opposite results, which has the highest transaction rate, about 40 K/min. But, the QoA measures are unsatisfactory, with the highest staleness, almost 50% of the CDC requests expiring. The results of the middle gear are between the high and low gears, with the τ (39 K/min) and the latency L (60 ms). However, the staleness S (30 ms delay) and the expiration ratio E (30%) are not acceptable. In contrast, the proposed workload-aware CDC outperforms the middle gear in terms of the S and the E, which are the same as in the high gear. The latency τ is the same as the middle gear, even though the L is slightly higher (80 ms). It is because the workload-aware scheme has taken into account the source-side impact, and exploited the deadlines of incoming CDC requests while delivering high delta freshness in time. Therefore, it can stop the deteriorating trend in performance if an appropriate gear is switched at runtime.

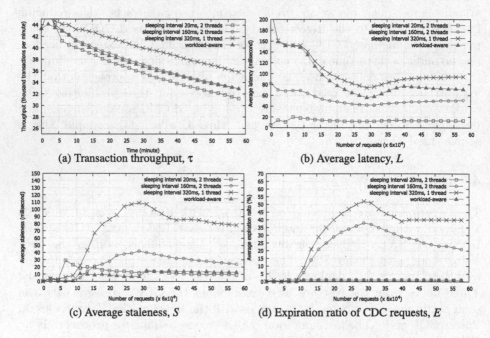

(a) Transaction throughput, τ (b) Average latency, L

(c) Average staleness, S (d) Expiration ratio of CDC requests, E

Fig. 3. QoT & QoA of trigger-based CDC variants

Evaluation of the Log-Based CDC. As described earlier, the log-based CDC cannot exploit worker parallelism due to sequential read of log files. We therefore only compare the CDC settings between two constant scheduling frequencies: *high gear* (sleeping interval 0 ms) and *low gear* (sleeping interval 120 ms) and a *workload-aware* scheme of changing the two frequencies. Figure 4 shows the results. Similar to the results of the trigger-based CDC, the log-based CDC processes in the high gear and the low gear show two extreme cases: the transaction throughput τ of low gear is better, while L, S and E of high gear are better. In contrast, the proposed workload-aware log-based CDC can achieve almost the same τ as the low gear while keeping most of the incoming CDC requests achieving the median staleness and expiration ratio. We see that the overall QoS can still benefit from the workload-aware approach without worker parallelism.

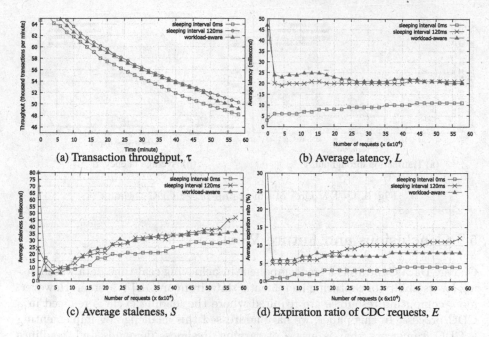

Fig. 4. QoT & QoA of log-based CDC variants

Evaluation of the Timestamp-Based CDC. Figure 5 shows the results for four timestamp-based CDC settings, *high gear* (sleeping interval 20 ms), *middle gear* (sleeping interval 640 ms), *low gear* (sleeping interval 1280 ms) and the proposed *workload-aware* setting with mixed gears. As shown, the transaction throughput τ under the workload-aware setting is very close to the highest throughput corresponding to the low speed gear, while the average staleness E remains as low as the high gear. These results correspond to the same conclusions drawn from the trigger-based and log-based CDC variants presented above. With the dynamic CDC settings customised at runtime, the performance gain in the workload-aware scheme outperforms the other three constant CDC settings throughout the experiment.

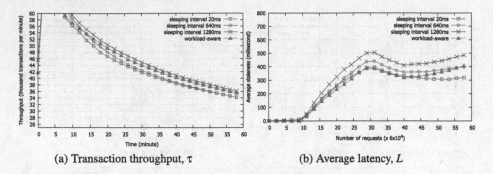

(a) Transaction throughput, τ (b) Average latency, L

Fig. 5. QoT & QoA of timestamp-based CDC variants

5 Conclusions and Future Work

Change Data Capture is a key component in achieving real-time data warehousing for its ability of capturing incremental data from an OLTP system. However, more computing resources are required where the loading cycle is reduced in a CDC process. In this paper, we have addressed this challenge by implementing a CDC framework that is aware of varying freshness demands and deadlines specified in the OLAP queries while meeting QoS requirements on the data source platforms. The proposed framework can adjust the so-called CDC gears at runtime according to the performance gain as the feedback from both OLTP and OLAP workload. We evaluated the framework and the results confirm that the proposed workload-aware CDC framework is capable of reacting to varying QoS requirements at runtime and balancing the overall QoS for both OLTP and OLAP workloads.

For future work, it would be interesting to make CDC a standardised component integrated into an ETL tool, to provide support for parallel processing, and to support beyond the workload aware, for example, on the basis of business criticality or data sensitivity.

Acknowledgement. This research was supported by the HEAT 4.0 project (8090-00046b) funded by Innovationsfonden.

References

1. Hamdi, I., Bouazizi, E., Alshomrani, S., Feki, J.: Improving QoS in real-time data warehouses by using feedback control scheduling. J. Inf. Deci. Sci. **10**(3), 181–211 (2018)
2. Kozielski, S., Wrembel, R. (eds): New Trends in Data Warehousing and Data Analysis. Springer (2009). https://doi.org/10.1007/978-0-387-87431-9
3. Liu, X.: Data warehousing technologies for large-scale and right-time data. Aalborg University, Defensed on June (2012)
4. Liu, X., Iftikhar, N.: An ETL optimization framework using partitioning and parallelization. In: Proceedings of the 30th SAC, pp. 1015–1022 (2015)

5. Shi, J., Guo, S., Luan, F., Sun, L.: Qos-ls: Qos-based load scheduling algorithm in real-time data warehouse. In: Proceedings of the 5th ICMMCT (2017)
6. Thiele, M., Fischer, U., Lehner, W.: Partition-based workload scheduling in living data warehouse environments. Inf. Syst. **34**(4–5), 382–399 (2009)
7. Thomsen, C., Pedersen, T.B., Lehner, W.: Rite: providing on-demand data for right-time data warehousing. In: Proceedings of the 24th ICDE, pp. 456–465 (2008)
8. Vassiliadis, P., Simitsis, A.: Near real time ETL. In: New Trends in Data Warehousing and Data Analysis, pp. 1–31. Springer (2009). https://doi.org/10.1007/978-0-387-87431-9

Machine Learning and Analtyics

Motif Based Feature Vectors: Towards a Homogeneous Data Representation for Cardiovascular Diseases Classification

Hanadi Aldosari[1]([✉]), Frans Coenen[1]([✉]), Gregory Y. H. Lip[2]([✉]),
and Yalin Zheng[3]([✉])

[1] Department of Computer Science, University of Liverpool, Liverpool L69 3BX, UK
{H.A.Aldosari,Coenen}@liverpool.ac.uk
[2] Liverpool Centre for Cardiovascular Science, University of Liverpool and Liverpool
Heart and Chest Hospital, Liverpool L69 7TX, UK
Gregory.Lip@liverpool.ac.uk
[3] Department of Eye and Vision Science, University of Liverpool,
Liverpool L7 8TX, UK
yalin.zheng@liverpool.ac.uk

Abstract. A process for generating a unifying motif-based homogeneous feature vector representation is described and evaluated. The motivation was to determine the viability of this representation as a unifying representation for heterogeneous data classification. The focus for the work was cardiovascular disease classification. The reported evaluation indicates that the proposed unifying representation is a viable one, producing better classification results than when a Recurrent Neural Network (RNNs) was applied to just ECG time series data.

Keywords: Motifs · Feature extraction and selection · Cardiovascular disease classification

1 Introduction

Time series classification is a common machine learning application domain [5–7]. Using the computer processing power that is now frequently available the size of the time series we wish to process has become less of a challenge. However, for many time series applications, the data is presented in a range of formats, not just time series. One example is Cardiovascular Disease (CVD) classification, where typically the data comprises electrocardiogram (ECGs), echocardiograms (Echo) and tabular patient data. Deep learners, such as RNNs, do not readily lend themselves to such heterogeneous data. One solution is to train a number of classifiers, one directed at each data format, and then combine the classification results. However, this assumes that the data sources are independent. The alternative is to adopt a unifying representation so that a single classification model can be generated. The most appropriate unifying representation, it is argued here, is a feature vector representation because this is compatible with a wide

M. Golfarelli et al. (Eds.): DaWaK 2021, LNCS 12925, pp. 235–241, 2021.
https://doi.org/10.1007/978-3-030-86534-4_22

range of classification models. The challenge is then how to extract appropriate features from data so as to construct the desired feature vector representation.

This paper explores the idea of generating a Homogeneous Feature Vector Representation (HFVR) from heterogeneous data by considering how we might extract features from time series that can be included in such a representation; such as ECG time series. The idea promoted in this paper is the use of motifs as time series features that can be coupled with other features in a HFVR. The challenge here is the computational complexity of finding exact motifs [4]. Recently the use of matrix profiles has been proposed [8]. The mechanism proposed in this paper is founded on matrix profiles, namely the Correct Matrix Profile (CMP) technique described in [1]. A further challenge, once a set of motifs has been identified, is to select a subset of these motifs to be included in the final HFVR. The criteria here is the effectiveness of the generated classification model, we want to choose motifs (features) that are good discriminators of class.

To evaluate the utility of the proposed unifying approach a cardiovascular diseases classification application was considered; more specifically, the binary classification of Atrial Fibrillation (AF), the most common cardiac rhythm disorder. For the evaluation the China Physiological Signal Challenge 2018 (CPSC2018) data was used [3]. A SVM classification model was then generated using the proposed unifying representation and the performance compared with that of a classification models built using only the original time series data (an RNN was used). The proposed HFVR approach, that allows the inclusion of features from heterogeneous data sources, outperformed the time series only approach even though only a small amount of additional information was incorporated into the HFVR. It is anticipated that when further features from other data sources, such as Echo data, are added the proposed approach will significantly outperform classifiers built using only a single data source.

2   Application Domain and Formalism

Atrial Fibrillation (AF) can be identified from range of tests, but ECG analysis is considered to be the most reliable [2]. An ECG indicates the electrical activity of the heart in terms of a summation wave that can be visualised and hence interpreted. A set of ECG records is of the form $\{R_1, R_2, \dots\}$ where each record R_i comprises a set of time series $\{T_1, T_2, \dots\}$ associated with a patient. The number of time series is usually 6 or 12 depending on whether six-lead or twelve-lead ECG has been used. For model training and evaluation we turn this data into an training/test data of the form $\mathbf{D} = \{\langle T_1, c_1\rangle, \langle T_2, c_2\rangle, \dots \langle t_n, c_n\rangle\}$ where c is a class label drawn from a set of classes C. For the evaluation presented later in this paper a binary classification scenario is presented, thus $C = \{true, false\}$. Each time series T_j comprises a sequence of data values $[t_1, t_2, \dots, t_n]$. A motif m is then a sub-sequence of a time series t_j. M is set of motifs extracted from $\mathbf{T}$, $M = \{m_1, m_2, \dots\}$. Not all the motifs in M will be good discriminators of class so we prune M to give M' and then M'', the set of attributes for our feature vector representation. The input set for the classification model generation thus consists of a set of vectors $\{V_1, V_2, \dots\}$ where each vector V_i comprises a set of occurrence count values m$\{v_1, v_2, \dots\}$, and a class label $c \in C$, such that there is a one-to-one correspondence between the values and M''.

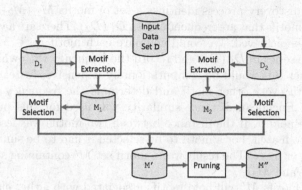

Fig. 1. Schematic of motif-based feature vector generation process.

3   Motif-Based Feature Vector Generation

This section presents the proposed motif feature extraction mechanism. A schematic describing the mechanism is given in Fig. 1. The input is the set **D** which is divided into D_1 and D_2, where D_1 corresponds to class c_1 and D_2 corresponds to class c_2. These are then processed to extract motifs. The motif extraction will result in two sets of motifs M_1 and M_2 corresponding to D_1 and D_2 respectively. We wish to identify motifs that occur frequently throughout D_1 and D_2 as these are assumed to be good indicators of class. The next stage is thus motif selection. This produces two refined sets of motifs $M_1' \subset M_1$ and $M_2' \subset M_2$ which are combined to form M'. We wish to identify motifs that are good discriminators of class thus we want to remove motifs from M' that are associated with both c_1 and c_2. The result is a pruned set of motifs, M'' which is then used to generate our HFVR.

The adopted approach for motif discovery was founded on the Guided Motif Search (GMS) algorithm [1]. The basic idea behind GMS is to discover meaningful motifs that represent the user's expected outcomes. That is achieved by combining the Matrix Profile (MP) with an Annotation Vector (AV) to produce a Correct Matrix Profile (CMP). The AV consists of real-valued numbers between 0 and 1 and has the same length as MP; there is a one-to-one correspondence between AV and MP. A high AV value indicates the sub-sequence at position i is a desirable motif. Each element cmp_i in the CMP is calculated using Eq. 1 where: (i) MP_i is the value in MP at position i, (ii) AV_i is the value from AV at position i, and (iii) $max(MP)$ denotes the maximum value in MP.

$$cmp_i = MP_i + ((1 - AV_i) \times max(MP)) \tag{1}$$

Thus if the $AV_i = 1$ the cmp_i value will be the same as the MP_i value, otherwise the cmp_i value will be the MP_i value increased by the $max(MP)$ value. The cmp_i value thus indicates whether position i contains a meaningful motif (to be selected), or not.

The motif discovery process identifies a set of motifs M_1 (M_2). However, we wish to retain motifs that are frequent across D_1 (D_2). There are a variety of ways that this can be achieved. We could compare each motif $m_i \in M_1$ $(m_j \in M_2)$ with each time series in D_1 (D_2) and record the frequency with which each motif occurs. However this would be computationally expensive. Instead we compare each motif with every other motif and determine the frequency of occurrence of each motif. Euclidean distance similarity was used for this purpose with a similarity threshold σ. If the distance between two motifs was less than σ they were deemed to match. For a motif to be selected it had to be similar to at least k other motifs or more. The result was a motif set M' containing selected motifs from both M_1 and M_2.

The set of motifs M' will hold motifs associated with either class c_1 or class c_2, and motifs associated with both classes c_1 and c_2. The motifs associated with both classes will not be good discriminators of class, hence these should be pruned. Thus, we compared each motif in class c_1 with every motif in class c_2. Euclidean distance with a σ similarity threshold was again used. If the Euclidean distance similarity was less than σ the motifs being compared were deemed to be representative of both class c_1 and c_2 and should therefore be excluded. The result is a pruned set of motifs, M'' which contains unique motifs with respect to classes c_1 and c_2 to be used in the desired HFVR.

The last stage in the proposed process was the feature vector generation stage. During this stage the set $\mathbf{V} = \{V_1, V_2, \dots\}$ was generated using the identified set of motifs M'' and any additional features we might wish to add (forb example age and gender). Each vector $V_i = \{v_1, v_2, \dots, c\} \in \mathbf{V}$ represents an ECG time series $T_i \in \mathbf{T}$. Each value v_j is either the numeric occurrence count of a motif m_j in the time series T_i, or the value associated with some additional feature. When classifying previously unseen records there will be no element c in V_i as this is the class value we wish to predict.

4   Evaluation

For the evaluation, The China Physiological Signal Challenge 2018 (CPSC2018) data set was used. For the proposed HFVR to be of value it needs to produce a classification accuracy that outperforms mechanisms founded on a single data source, such as only ECG time series. A SVM classification model was used with respect to the proposed HFVR. An RNN was used with respect to the time series only classification. The metrics used for the evaluation were accuracy, precision, recall and F1. Ten fold cross-validation was used throughout. The objectives of the evaluation were: (i) to identify the most appropriate value for σ, (ii) to identify the most appropriate value for k, and (iii) to determine the effectiveness of the proposed motif-based feature vector approach in comparison with an a deep learner applied directly to the input ECG time series.

To determine the most appropriate value for σ, a sequence of experiments was conducted using a range of values for σ from 0.05 to 0.50 increasing in steps of 0.05. The value of k used was 150 because preliminary experiments indicated this to be an appropriate value. The best recorded F1 value was 73.33%, obtained using $\sigma = 0.15$. This was thus the value for σ used for the further experiments reported on in the following sub-sections.

To determine the most appropriate value for k, a further sequence of experiments was conducted using a range of values for k from 50 to 250 increasing in steps of 20. The results obtained indicated that the best F1 value of 75.05% was obtained using $k = 90$. This was then the k used with respect to the additional experiments reported on below.

To ascertain whether the proposed HFVR operated in an effective manner a SVM classification model was generated and tested using the proposed representation and compared to a RNN generated from just the ECG time series data. Experiments were conducted using the SVM with fixed parameters and with GridSearch to identify best parameters. For the RNN the SimpleRNN algorithm was used (available as part of the Keras open-source software Python library), where the output for each time stamp layer is fed to next time stamp layer. SimpleRNN was applied directly to the time series data. With respect to the proposed representation experiments were conducted using: (i) just motifs, (ii) motifs + gender, (iii) motifs + age, and (iv) motifs + gender + age.

Table 1. Comparison of proposed approach using SVM without GridSearch and RNN.

	Accuracy	Precision	Recall	F1
Proposed approached (motifs)	76.68%	72.96%	78.93%	74.32%
Proposed approached (motifs + gender)	77.61%	72.86%	80.58%	75.71%
Proposed approached (motifs + age)	85.09%	86.22%	84.71%	85.16%
Proposed approached (motifs + gender+age)	**85.28%**	**86.26%**	84.82%	**85.28%**
SimpleRNN model	81.17%	85.30%	75.33%	79.00%

The results using SVM are given in Tables 1 and 2; best results highlighted in bold font. From the Table it can firstly be observed that motifs when used on their own do not perform as well as when an RNN is applied to the time series data directly. Secondly it can be observed that the inclusion of gender to the feature vector representation does not make a significant difference, while including age does make a significant difference. Using motifs and age, or motifs, age and gender, produces a performance better than the time series only RNN performance. Age is the most significant factor here; it is well known that AF is more prevalent in older age groups. Thirdly, using GridSearch to identify best parameters for SVM classification also serves to enhance performance. Fourthly that best results are obtained when using GridSearch and a feature vector that includes motifs + age. Finally, from the results, an argument can be made that gender acts as a confounder, in that when GridSearch is used with feature vectors made up of motifs + age + gender the results are not as good as when using feature vectors made up of motifs + age.

Table 2. Comparison of proposed approach using SVM with GridSearch and RNN.

	Accuracy	Precision	Recall	F1
Proposed approached (motifs)	79.57%	78.51%	79.45%	78.01%
Proposed approached (motifs + gender)	80.17%	77.52%	82.0%	79.22%
Proposed approached (motifs + age)	**86.41%**	**87.27%**	**86.18%**	**86.37%**
Proposed approached (motifs + gender+age)	85.49%	86.22%	85.26%	85.47%
SimpleRNN model	81.17%	85.30%	75.33%	79.00%

5   Conclusion

A mechanism for generating a motif-based feature vector representation for use with CVD classification has beern presented. The motivation was that effective CVD classification requires input from a number of sources (typically ECG, Echo and tabular data) and that a feature vector representation would provide a unifying mechanism for representing such heterogeneous data. The concept was evaluated by comparing its operation, using motifs paired with age and/or gender data, in the context of CVD classification, with that when using a RNN applied on to the ECG time series alone. The most appropriate values for σ and k were found to be 0.15 and 90 respectively. The classification results produced indicated that using the proposed representation combing motifs + age, or motifs + age + gender, produced a better classification than when using the time series on their own; thus indicating the benefits of the proposed unifying representation.

References

1. Dau, H.A., Keogh, E.: Matrix profile V: a generic technique to incorporate domain knowledge into motif discovery. In: Proceedings of the 23rd ACM SIGKDD International Conference on Knowledge Discovery and Data Mining, pp. 125–134 (2017)
2. Lip, G., et al.: Atrial fibrillation. Nat. Rev. Dis. Primers **31**, 16016 (2016)
3. Liu, F., et al.: An open access database for evaluating the algorithms of electrocardiogram rhythm and morphology abnormality detection. J. Med. Imaging Health Inform. **8**(7), 1368–1373 (2018)
4. Mueen, A., Keogh, E., Zhu, Q., Sydney Cash, S., Westover, B.: Exact discovery of time series motifs. In: SIAM International Conference on Data Mining, pp. 473–484. SIAM (2009)
5. Oh, S.L., Ng, E.Y., San Tan, R., Acharya, U.R.: Automated diagnosis of arrhythmia using combination of CNN and LSTM techniques with variable length heart beats. Comput. Biol. Med. **102**, 278–287 (2018)

6. Pourbabaee, B., Roshtkhari, M.J., Khorasani, K.: Deep convolutional neural networks and learning ECG features for screening paroxysmal atrial fibrillation patients. IEEE Tran. Syst. Man Cybern. Syst. **48**(12), 2095–2104 (2018)
7. Wang, G., et al.: A global and updatable ECG beat classification system based on recurrent neural networks and active learning. Inf. Sci. **501**, 523–542 (2019)
8. Yeh, C.C.M., et al.: Matrix profile I: all pairs similarity joins for time series: a unifying view that includes motifs, discords and shapelets. In: IEEE 16th International Conference on Data Mining (ICDM), pp. 1317–1322. IEEE (2016)

Filter-Based Feature Selection Methods for Industrial Sensor Data: A Review

Sabrina Luftensteiner[✉], Michael Mayr, and Georgios Chasparis

Software Competence Center Hagenberg GmbH, Hagenberg im Muehlkreis, Austria
{sabrina.luftensteiner,michael.mayr,georgios.chasparis}@scch.at
http://www.scch.at/

Abstract. The amount of produced data in industrial environments is continuously rising, containing relevant but also irrelevant or redundant data. The size and dimensionality of gathered data makes it difficult to identify features that have a high influence on a given target using domain information alone. Therefore, some kind of dimensionality reduction is necessary to extract the most important features. To provide an explainable feature selection process and avoid the creation of less explainable features, filter feature selection methods provide good approaches. Existing surveys cover a broad range of methods, however the provided comparative analysis is often restricted to a limited set of methods or to a limited number of datasets. This paper will cover a broad range of filter feature selection methods and compare them in three scenarios, ranging from low to high dimensional data, to determine their usability in varying settings.

Keywords: Feature selection · Multi-variate data · Sensor data

1 Introduction

1.1 Background

Industrial digitization is expanding with a relatively fast pace, given that sensors are getting continuously faster, better and cheaper [35]. For the control of processes in process industry, the most important features have to be identified, facing the problem of variable feature space dimensions as well as noisy, irrelevant, redundant and duplicated features [27]. The identification and extraction of relevant features enable process operators to monitor and control selected vital

This paper is supported by European Union's Horizon 2020 research and innovation programme under grant agreement No 869931, project COGNIPLANT (COGNITIVE PLATFORM TO ENHANCE 360° PERFORMANCE AND SUSTAINABILITY OF THE EUROPEAN PROCESS INDUSTRY). It has also been supported by the Austrian Ministry for Transport, Innovation and Technology, the Federal Ministry of Science, Research and Economy, and the Province of Upper Austria in the frame of the COMET center SCCH.

M. Golfarelli et al. (Eds.): DaWaK 2021, LNCS 12925, pp. 242–249, 2021.
https://doi.org/10.1007/978-3-030-86534-4_23

process parameters instead of all parameters and additionally allow the efficient training of machine learning models or further statistical evaluations [27]. Especially, machine learning models have difficulty in dealing with high dimensional input features [29] and suffer easily the curse of dimensionality [28], resulting in computationally expensive model training sessions as well as overfitting and biased models. The integration of feature selection as an additional data preprocessing step enables the advantages of better model interpretability, better generalization of the model, reduction of overfitting and efficiency in terms of training time and space complexity [21]. Feature selection methods essentially divide into filter, wrapper and embedded methods. Filter methods select subsets of variables or rank them as preprocessing step, completely independent of the chosen predictor method. Wrapper methods utilize a black box to score subsets of features according to their predictive power and have a high time complexity due to their brute force nature [21]. Embedded methods are integrated into the training process and are specifically designed for learning approaches [24], e.g. decision trees or support vectors. Filter methods are working independently of machine learning algorithms and score each feature individually, not projecting them on a lower feature space [21], and avoid overfitting [24]. The methods use mainly individual evaluation approaches and create ranked based outputs [9].

In this paper, we present various filter feature selection methods and compare the most promising in experimental settings for low, medium and high dimensional datasets based on their performance on multiple levels. The overall goal of this paper is to provide a stable base for feature selection in industrial settings.

1.2 Previous Surveys

Many feature selection methods have been proposed in literature and their comparative study might be a challenging task. Various surveys cover feature selection methods tailored for classification tasks, including [8,13] and [29]. An overview on feature selection methods for clustering tasks is covered in [14] and [29]. Goswami et al. cover a broad base for feature selection in their survey, including the processing of discrete, continuous, binary and ordinary data [21].

Most of the these surveys do not mention regression tasks, which are very important in industrial scenarios. Furthermore, the methods were primarily tested and compared on datasets with equal dimensions, avoiding comparisons between low, medium and high dimensional data. This paper deals with these missings.

2   Supervised Feature Selection Methods

The following section is dedicated to supervised feature selection methods. The selected methods cover proven approaches and adaptations of such approaches, e.g. if methods are based upon categorical data instead of numerical data. The methods are partitioned, according to [13], into four categories: dependence measures, distance measures, information measures and consistency measures. Methods may fit into multiple categories, in which case the most suitable is selected.

Dependence Measures. Dependence measures are often referred to as correlation measures and incorporate the dependence between two or more features to indicate the degree their redundancy [13]. Pearson's and Spearman correlation coefficients represent straightforward methods, which provide ranked features as results. Pearson's Correlation Coefficient (PCC) [5] detects linear relations between numeric features and works in uni-variate settings [21]. Spearman correlation coefficients (SCC) evaluate if a monotonic relationship between two continuous or ordinal variables is available [37]. Haindl et al. proposed a new mutual correlation (MC) approach based upon an adapted correlation calculation for fast and accurate results with high dimensional data [25]. Correlation-based feature selection (CFS) is based on the fact that the most important features are correlated with the target and less with each other [26] and supports multi-variate settings with all types of data [21]. The fast correlation-based filter (FCBF) is a fast multi-variate filter method that can identify relevant features as well as redundancy among relevant features without pairwise correlation analysis [43].

A method called Minimum Redundancy - Maximum Relevance (MRMR) was presented by Ding et al. and reduces mutual redundancy within a feature set to capture the class characteristics in a broader scope. The method is applicable for classification and regression tasks and supports multi-variate settings [17].

The Hilbert-Schmidt Independence Criterion (HSIC) [22] is often used as a measure of dependence between the features and the target, e.g. for the methods BAHSIC and FOHSIC [38,39]. Depending on the used kernels, the method is applied to regression or classification tasks in multi-variate settings. Kernel-based methods are particularly well suited for high dimensional data and a low amount of samples. For higher numbers of samples, other methods, such as Mutual Information (MI), are computationally more efficient [7]. Chang et al. propose a method, which is robust to different relationship types of features and robust to unequal sample sizes of different features [10]. The method supports multi-variate settings but is computationally expensive for high dimensional datasets and therefore less suitable for industrial use cases. The filter method ReliefF detects conditional dependencies between attributes using neighborhood search [36]. MICC combines Mutual Information (MI) and Pearson Correlation Coefficient [23]. T-statistics, ANOVA, F-Test and Chi-square based filter methods are also often used for feature selection [8,21]. Methods developed for high-dimensional data are provided by Yu et al. [43] and Chormunge et al. [12].

Distance Measures. Distance measures are also known as divergence measures and are utilized more often for feature selection in classification scenarios [13]. This type of ranked filter methods calculates the class separability of each feature, where numeric values have to be discretized [8]. Methods are covering Kulback-Leibler divergence [40], Bayesian distance [11], Bhattacharyya coefficient [16], Kolmogorov variational distance [44] and the Matusita distance [6]. Other methods in this category are often wrapper methods, e.g. the Firefly optimization algorithm by Emary et al. [18] or the invariant optimal feature selection with distance discriminants and feature ranking by Liang et al. [31].

Information Measures. Measures of this category are used to calculate the ranked information gain from features [13]. Mutual information (MI) measures the amount of information that one random variable has about another variable [41]. CIFE is a filter method that maximizes the joint class-relevant information by explicitly reducing the class-relevant redundancies among features [32]. CMIM maximizes the mutual information of features with the class to predict conditional to any feature already picked. It ensures the selection of features that are both individually informative and two-by-two weakly dependent [20]. DISR uses as a criterion a set of variables that can return information on the output class that is higher than the sum of the information of each variable taken individually [34]. JMI finds the optimal viewing coordinates with the highest joint mutual information and eliminates redundancy in input features [42]. MIFS takes both the mutual information with respect to the output class and with respect to the already selected features into account [4]. MIM uses the maximization of mutual information [30] and NMIFS uses normalized mutual information [19].

Consistency Measures. Consistency measures are used to find minimally sized subsets that satisfy an acceptable inconsistency rate [13,15]. FOCUS [1], FOCUS2 [2], CFOCUS [3] and Liu's consistency measure [33] are often used to calculate these inconsistency rates and generate feature subsets.

3 Experiments and Discussion

This section covers a comparison of feature selection methods mentioned in Sect. 2 regarding their speed, resource allocation, amount of selected features and the accuracy on a machine learning model. The experiments are split into low (81 columns × 21263 rows), medium (5380 columns × 2205 rows) and high dimensional (43680 columns × 2205 rows) scenarios. Methods, which provide ranked features, use a predefined amount of features for the evaluation of the machine learning model depending on the highest amount of a subset feature selection method to enable fair comparisons. The hardware used for the experiments covers an Intel(R) Core(TM) i7-8665U CPU @ 1.9 GHz and 32 GB RAM. The machine learning part uses the selected features to train a random forest using 10-Fold cross validation, using the mean R2 value as accuracy of the model.

Low Dimensional Data. The low dimensional data scenario covers a use case regarding superconductivity[1] in an industrial environment. The features are entirely numerical and multi-variate. The goal is to predict the critical temperature of a superconductor. Table 1 contains the results for the experiment, whereat the baseline R2 is 0.95. Most feature selection methods are able to exceed this baseline with exception of CIFE. The resource allocation is relatively evenly low, whereat FCBF needs slightly more resources compared to other methods. The duration of selecting/ranking features is relatively low. MICC, MIM, DISR and CFS needed more than a minute and were removed from the table.

---

[1] https://archive.ics.uci.edu/ml/datasets/Superconductivty+Data#.

Table 1. Results of the low dimensional use-case (21263 samples × 81 columns). The baseline random forest model without feature selection has a R2 score of 0.95.

Method	Duration (sec)	Resources allocation	# Sel. features	R2 RF	Category
FCBF	45	low/medium	3	**0.9998**	Dependence
F-Test	6	low	7	0.9977	Dependence
ReliefC	5	low	10 (ranked)	**0.9997**	Dependence
IG	5	low	10 (ranked)	0.9849	Information
HSIC	10	low	10 (ranked)	0.9936	Dependence
MI	11	low	10 (ranked)	**0.9994**	Information
CIFE	19	low	10 (ranked)	0.9453	Information
PCC	21	low	10 (ranked)	0.9915	Dependence
SCC	22	low	10 (ranked)	0.9918	Dependence
CMIM	53	low	10 (ranked)	0.9581	Information
JMI	57	low	10 (ranked)	0.9941	Information
MRMR	58	low	10 (ranked)	**0.9991**	Information
MIFS	60	low	10 (ranked)	0.9937	Information

Medium Dimensional Data. The medium dimensional data scenario covers a subset of sensor records regarding the monitoring of hydraulic systems[2] in an industrial environment. It is numerical and multi-variate. Table 2 contains the results for the second experiment, whereat the baseline R2 is 0.97. Durations above 15 min were removed and affect PCC, SCC, MRMR, MIM and MIFS. The resource allocation is higher for more complex approaches.

Table 2. Results of the medium dimensional use-case (2205 samples × 5380 columns). The baseline random forest model without feature selection has a R2 score of 0.97.

Method	Duration (min)	Resources allocation	# Sel. features	R2 RF	Category
FCBF	5	low/medium	5	**0.9955**	Dependence
F-Test	1	low	9	**0.9951**	Dependence
HSIC	1	low	10 (ranked)	0.9180	Dependence
ReliefC	1	low	10 (ranked)	**0.9972**	Dependence
IG	1	low	10 (ranked)	0.9907	Information
MI	3	low/medium	10 (ranked)	**0.9983**	Information
CIFE	10	low/medium	10 (ranked)	0.9923	Information
CMIM	10	low/medium	10 (ranked)	0.9916	Information
JMI ·	12	low/medium	10 (ranked)	0.9942	Information

High Dimensional Data. The high dimensional data scenario covers the entire dataset of sensor records regarding the monitoring of hydraulic systems (see Footnote 2) in an industrial environment. It consists numerical features and is

---

[2] https://archive.ics.uci.edu/ml/datasets/Condition+monitoring+of+hydraulic+systems.

multi-variate. Table 3 contains the results for the third experiment, whereat baseline R2 is 0.94. Due to taking more than 60 min, CIFE, CMIM, JMI, MIFS, MIM and MRMR are removed. The remaining methods mostly performed better than the baseline and are recommandable for usage in high dimensional scenarios.

Table 3. Results of the high dimensional use-case (2205 samples × 43680 columns). The baseline random forest model without feature selection has a R2 score of 0.93.

Method	Duration (min)	Resources allocation	# Sel. features	R2 RF	Category
FCBF	22	medium	12	**0.9979**	Dependence
HSIC	23	medium	7	0.8963	Dependence
F-Test	2	low/medium	12 (ranked)	0.9874	Dependence
ReliefC	2	low/medium	12 (ranked)	**0.9920**	Dependence
IG	2	low/medium	12 (ranked)	0.9619	Information
MI	19	low/medium	12 (ranked)	**0.9952**	Information

4 Conclusion

Due to the rising amount of gathered data in industry, feature selection gets more important to extract relevant data and remove irrelevant, noisy and redundant data for further processing. To provide an explainable feature selection process and avoid the creation of less explainable features, filter feature selection methods provide promising approaches. This paper gives an overview of various types of filter feature selection methods and compares them in low, medium and high dimensional use-case scenarios regarding their duration and accuracy improvements in machine learning models. In our experiments, FCBF, ReliefC and MI appear that they are working well with varying data dimensions and select favorable features for further usage, e.g. model training.

References

1. Almuallim, H., Dietterich, T.G.: Learning with many irrelevant features. In: AAAI, vol. 91, pp. 547–552. Citeseer (1991)
2. Almuallim, H., Dietterich, T.G.: Learning Boolean concepts in the presence of many irrelevant features. Artif. Intell. **69**(1–2), 279–305 (1994)
3. Arauzo-Azofra, A., Benitez, J.M., Castro, J.L.: Consistency measures for feature selection. J. Intell. Inf. Syst. **30**(3), 273–292 (2008)
4. Battiti, R.: Using mutual information for selecting features in supervised neural net learning. IEEE Trans. Neural Networks **5**(4), 537–550 (1994)
5. Benesty, J., Chen, J., Huang, Y., Cohen, I.: Pearson correlation coefficient. In: Noise Reduction in Speech Processing. STSP, vol. 2, pp. 1–4. Springer, Heidelberg (2009). https://doi.org/10.1007/978-3-642-00296-0_5
6. Bruzzone, L., Roli, F., Serpico, S.B.: An extension of the Jeffreys-Matusita distance to multiclass cases for feature selection. IEEE Trans. Geosci. Remote Sens. **33**(6), 1318–1321 (1995)

7. Camps-Valls, G., Mooij, J., Scholkopf, B.: Remote sensing feature selection by kernel dependence measures. IEEE Geosci. Remote Sens. Lett. **7**(3), 587–591 (2010)
8. Cantú-Paz, E., Newsam, S., Kamath, C.: Feature selection in scientific applications. In: Proceedings of the Tenth ACM SIGKDD International Conference on Knowledge Discovery and Data Mining, pp. 788–793 (2004)
9. Chandrashekar, G., Sahin, F.: A survey on feature selection methods. Comput. Electr. Eng. **40**(1), 16–28 (2014)
10. Chang, Y., Li, Y., Ding, A., Dy, J.: A robust-equitable copula dependence measure for feature selection. In: Artificial Intelligence and Statistics, pp. 84–92. PMLR (2016)
11. Chen, C.: On information and distance measures, error bounds, and feature selection. Inf. Sci. **10**(2), 159–173 (1976)
12. Chormunge, S., Jena, S.: Correlation based feature selection with clustering for high dimensional data. J. Electr. Syst. Inf. Technol. **5**(3), 542–549 (2018)
13. Dash, M., Liu, H.: Feature selection for classification. Intell. Data Anal. **1**(1–4), 131–156 (1997)
14. Dash, M., Liu, H.: Feature selection for clustering. In: Terano, T., Liu, H., Chen, A.L.P. (eds.) PAKDD 2000. LNCS (LNAI), vol. 1805, pp. 110–121. Springer, Heidelberg (2000). https://doi.org/10.1007/3-540-45571-X_13
15. Dash, M., Liu, H.: Consistency-based search in feature selection. Artif. Intell. **151**(1–2), 155–176 (2003)
16. Derpanis, K.G.: The Bhattacharyya measure. Mendeley. Computer **1**(4), 1990–1992 (2008)
17. Ding, C., Peng, H.: Minimum redundancy feature selection from microarray gene expression data. J. Bioinform. Comput. Biol. **3**(02), 185–205 (2005)
18. Emary, E., Zawbaa, H.M., Ghany, K.K.A., Hassanien, A.E., Parv, B.: Firefly optimization algorithm for feature selection. In: Proceedings of the 7th Balkan Conference on Informatics Conference, pp. 1–7 (2015)
19. Estévez, P.A., Tesmer, M., Perez, C.A., Zurada, J.M.: Normalized mutual information feature selection. IEEE Trans. Neural Networks **20**(2), 189–201 (2009)
20. Fleuret, F.: Fast binary feature selection with conditional mutual information. J. Mach. Learn. Res. **5**(9) (2004)
21. Goswami, S., Chakrabarti, A.: Feature selection: a practitioner view. Int. J. Inf. Technol. Comput. Sci. (IJITCS) **6**(11), 66 (2014)
22. Gretton, A., Bousquet, O., Smola, A., Schölkopf, B.: Measuring statistical dependence with Hilbert-Schmidt norms. In: Jain, S., Simon, H.U., Tomita, E. (eds.) ALT 2005. LNCS (LNAI), vol. 3734, pp. 63–77. Springer, Heidelberg (2005). https://doi.org/10.1007/11564089_7
23. Guha, R., Ghosh, K.K., Bhowmik, S., Sarkar, R.: Mutually informed correlation coefficient (MICC)-a new filter based feature selection method. In: 2020 IEEE Calcutta Conference (CALCON), pp. 54–58. IEEE (2020)
24. Guyon, I., Elisseeff, A.: An introduction to variable and feature selection. J. Mach. Learn. Res. **3**, 1157–1182 (2003)
25. Haindl, M., Somol, P., Ververidis, D., Kotropoulos, C.: Feature selection based on mutual correlation. In: Martínez-Trinidad, J.F., Carrasco Ochoa, J.A., Kittler, J. (eds.) CIARP 2006. LNCS, vol. 4225, pp. 569–577. Springer, Heidelberg (2006). https://doi.org/10.1007/11892755_59
26. Hall, M.A.: Correlation-based feature selection for machine learning (1999)
27. Jeong, B., Cho, H.: Feature selection techniques and comparative studies for large-scale manufacturing processes. Int. J. Adv. Manuf. Technol. **28**(9), 1006–1011 (2006)

28. Koller, D., Sahami, M.: Toward optimal feature selection. Technical Report, Stanford InfoLab (1996)
29. Kumar, V., Minz, S.: Feature selection: a literature review. SmartCR 4(3), 211–229 (2014)
30. Lewis, D.D.: Feature selection and feature extract ion for text categorization. In: Speech and Natural Language: Proceedings of a Workshop Held at Harriman, 23–26 February, 1992, New York (1992)
31. Liang, J., Yang, S., Winstanley, A.: Invariant optimal feature selection: a distance discriminant and feature ranking based solution. Pattern Recogn. 41(5), 1429–1439 (2008)
32. Lin, D., Tang, X.: Conditional infomax learning: an integrated framework for feature extraction and fusion. In: Leonardis, A., Bischof, H., Pinz, A. (eds.) ECCV 2006. LNCS, vol. 3951, pp. 68–82. Springer, Heidelberg (2006). https://doi.org/10.1007/11744023_6
33. Liul, H., Motoda, H., Dash, M.: A monotonic measure for optimal feature selection. In: Nédellec, C., Rouveirol, C. (eds.) ECML 1998. LNCS, vol. 1398, pp. 101–106. Springer, Heidelberg (1998). https://doi.org/10.1007/BFb0026678
34. Meyer, P.E., Bontempi, G.: On the use of variable complementarity for feature selection in cancer classification. In: Rothlauf, F., et al. (eds.) EvoWorkshops 2006. LNCS, vol. 3907, pp. 91–102. Springer, Heidelberg (2006). https://doi.org/10.1007/11732242_9
35. Reis, M.S., Gins, G.: Industrial process monitoring in the big data/industry 4.0 era: from detection, to diagnosis, to prognosis. Processes 5(3), 35 (2017)
36. Robnik-Šikonja, M., Kononenko, I.: Theoretical and empirical analysis of relieff and rrelieff. Mach. Learn. 53(1), 23–69 (2003)
37. Saeys, Y., Abeel, T., Van de Peer, Y.: Robust feature selection using ensemble feature selection techniques. In: Daelemans, W., Goethals, B., Morik, K. (eds.) ECML PKDD 2008. LNCS (LNAI), vol. 5212, pp. 313–325. Springer, Heidelberg (2008). https://doi.org/10.1007/978-3-540-87481-2_21
38. Song, L., Smola, A., Gretton, A., Bedo, J., Borgwardt, K.: Feature selection via dependence maximization. J. Mach. Learn. Res. 13(5) (2012)
39. Song, L., Smola, A., Gretton, A., Borgwardt, K.M., Bedo, J.: Supervised feature selection via dependence estimation. In: Proceedings of the 24th International Conference on Machine Learning, pp. 823–830 (2007)
40. Van Erven, T., Harremos, P.: Rényi divergence and Kullback-Leibler divergence. IEEE Trans. Inf. Theory 60(7), 3797–3820 (2014)
41. Vergara, J.R., Estévez, P.A.: A review of feature selection methods based on mutual information. Neural Comput. Appl. 24(1), 175–186 (2013). https://doi.org/10.1007/s00521-013-1368-0
42. Yang, H., Moody, J.: Feature selection based on joint mutual information. In: Proceedings of International ICSC Symposium on Advances in Intelligent Data Analysis, vol. 1999, pp. 22–25. Citeseer (1999)
43. Yu, L., Liu, H.: Feature selection for high-dimensional data: a fast correlation-based filter solution. In: Proceedings of the 20th International Conference on Machine Learning (ICML-03), pp. 856–863 (2003)
44. Zhang, Z.: Estimating mutual information via Kolmogorov distance. IEEE Trans. Inf. Theory 50(9), 3280–3282 (2007)

A Declarative Framework for Mining
Top-k High Utility Itemsets

Amel Hidouri[1,2]([envelope]), Said Jabbour[2], Badran Raddaoui[4], Mouna Chebbah[3],
and Boutheina Ben Yaghlane[1]

[1] LARODEC, University of Tunis, Tunis, Tunisia
boutheina.yaghlane@ihec.rnu.tn
[2] CRIL - CNRS UMR 8188, University of Artois, Lens, France
{hidouri,jabbour}@cril.fr
[3] LARODEC, Univ. Manouba, ESEN, Manouba, Tunisia
mouna.chebbah@esen.tn
[4] SAMOVAR, Télécom SudParis, Institut Polytechnique de Paris, Paris, France
badran.raddaoui@telecom-sudparis.eu

Abstract. The problem of mining high utility itemsets entails identifying a set of items that yield the highest utility values based on a given user utility threshold. In this paper, we utilize propositional satisfiability to model the Top-k high utility itemset problem as the computation of models of CNF formulas. To achieve our goal, we use a decomposition technique to improve our method's scalability by deriving small and independent sub-problems to capture the Top-k high utility itemsets. Through empirical evaluations, we demonstrate that our approach is competitive to the state-of-the-art specialized algorithms.

Keywords: Top-k · High utility · Propositional satisfiabilty

1 Introduction

Mining High Utility Itemsets (HUIM) as a major keystone in pattern discovery generalizes the problem of frequent itemsets (FIM). HUIM considers item weights to identify the set of items that appear together in a given transaction database and have a high importance to the user by using a *utility function*. An itemset is a *high utility itemset* (HUI) if its utility value is greater than a user threshold; otherwise, the itemset is a *low utility itemset*.

It is still difficult in practice for users to set an appropriate threshold value to find the entire set of high utility itemsets. In other words, a low threshold may entail a large number of patterns, whereas a high threshold value may give rise an empty set of HUIs. To handle this limitation, the Top-k high utility itemset mining (Top-k HUIM) is introduced. The user only needs to specify the desired number k of HUIs. Several methods have been studied to tackle this problem: the *two phase* based algorithms [4,5] in which computing the Top-k HUIs is done

© Springer Nature Switzerland AG 2021
M. Golfarelli et al. (Eds.): DaWaK 2021, LNCS 12925, pp. 250–256, 2021.
https://doi.org/10.1007/978-3-030-86534-4_24

in two steps. In contrast, *one-phase* methods [6–8] aim to compute the entire set of Top-k HUIs in a single phase without candidate generation.

In this paper, we introduce SATTHUIM (SAT based Top-k High Utility Itemset Mining), an approach for efficiently mining Top-k high utility itemsets embedded in a transaction database. This algorithm makes original use of propositional satisfiability. Our main contribution is a method based on propositional satisfiability to model and solve the task of mining the set of all Top-k HUIs. Then, we employ a decomposition technique that allows us to derive smaller and independent enumeration sub-problems to deal with the scalability issue.

2 Preliminaries

Let Ω denote a finite set of items. A transaction database $D = \{T_1, T_2, \ldots, T_m\}$ is a set of m transactions such that each transaction T_i is composed of a set of items, $T_i \subseteq \Omega$, s.t. T_i has a unique identifier i called its transaction identifier (TID). Each item $a \in \Omega$ has an external utility $w_{ext}(a)$ and an internal utility $w_{int}(a, T_i)$ in each transaction T_i.

Definition 1. *Given a transaction database D, the utility of an item a in a transaction $(i, T_i) \in D$ is $u(a, T_i) = w_{int}(a) \times w_{ext}(a, T_i)$. Then, the itemset's utility X in a transaction $(u(X, T_i))$, the utility of an itemset X in D $(u(X))$, the transaction utility of the transaction T_i in D $(TU(T_i))$, and the transaction weighted utilization of X $(TWU(X))$, are defined as follows:*

$$u(X, T_i) = \sum_{a \in X} u(a, T_i) \quad (1) \qquad u(X) = \sum_{(i,T_i) \in D \ |\ X \subseteq T_i} u(X, T_i) \quad (2)$$

$$TU(T_i) = \sum_{a \in T_i} u(a, T_i) \quad (3) \qquad TWU(X, D) = \sum_{(i,T_i) \in D \ |\ X \subseteq T_i} TU(T_i) \quad (4)$$

Definition 2. *Given a transaction database D and an itemset X. Then, X is a high utility closed itemset if there exists no other itemset X' s.t. $X \subset X'$, and $\forall (i, T_i) \in D$, if $X \in T_i$ then $X' \in T_i$. In addition, X is called a Top-k high utility itemset if there are less than k high utility itemsets of D with utilities larger than $u(X)$.*

A Conjunctive Normal Formula (CNF) Φ is defined over a set of propositional variables denoted $Var(\Phi) = \{x_1, \ldots, x_n\}$. Φ consists of a conjunction of clauses, where each clause c is a disjunction of literals. A literal is either a variable x_i or its complement $\neg x_i$. A Boolean interpretation Δ of a formula Φ is defined as a function from $\Lambda(\Phi)$ to $[0, 1]$ (1 corresponds to *true* and 0 to *false*). A model of a formula Φ is an interpretation Δ that satisfies Φ, i.e., $\Delta(\Phi) = 1$. The formula Φ is satisfiable if it has a model. In the sequel, $\models$ refers to the logical inference and $\models_*$ the one restricted to unit propagation.

3   Computing Top-k High Utility Itemsets Using SAT

In this section, we first review the formulation of HUIM into propositional satisfiability [2]. The proposed encoding consists of a set of propositional variables
to represent both items and transactions in the transaction database D. Each
item a (resp. transaction identifier i) is associated with a propositional variable,
referred to as p_a (resp. q_i). The encoding of HUIM into SAT is obtained through
a set of constraints as summarized in Fig. 1.

$$\bigwedge_{i=1}^{m}(\neg q_i \leftrightarrow \bigvee_{a \in \Omega \setminus T_i} p_a) \qquad (5) \qquad \Phi_{clos} = \bigwedge_{a \in \Omega}(p_a \vee \bigvee_{a \notin T_i} q_i) \qquad (6)$$

$$\sum_{i=1}^{m} \sum_{a \in T_i} u(a, T_i) \times (p_a \wedge q_i) \geq \theta \qquad (7)$$

Fig. 1. SAT Encoding Scheme for HUIM

Constraint (5) encodes the cover of the candidate itemset. This means that
the itemset is not supported by the ith transaction (i.e., q_i is *false*) iff an item not
appearing in the transaction i is set to *true*. The linear inequality (7) expresses
the utility constraint of an itemset X in D using a threshold θ. Moreover, the
propositional formula (6) selects the set of closed HUIs. It ensures that if all
transactions where the candidate itemset X is involved contain the item a, then
a must appear in X. As shown in [2], the CNF formula (5) $\wedge$ (7) encodes the
HUIM problem, while (5) $\wedge$ (7) $\wedge$ (6) models the closed HUIM task.

Now, we demonstrate how to use propositional satisfiability to find Top-k
HUIs in D. First, a set of k HUIs is generated. The utility of the kth itemset is
used to dynamically set the minimum utility threshold. Hence, each time a novel
itemset is found, the bound θ is fixed accordingly, and so on. Last, a SAT solver
is used as an oracle to compute all the models satisfying the required constraints.

To enumerate all models of the HUIM encoding, we employ an extension
of the procedure of Davis–Putnam, Logemann and Loveland (DPLL) (see Algorithm 1). In this procedure, a non assigned variable p of the formula Φ is selected,
and the algorithm extends the current interpretation by assigning p to *true*. Next,
unit propagation is performed (lines 2–3). If all literals are assigned without con-
flict, then Δ is selected as a model of the CNF formula (line 8–9). The procedure
is repeated until a set of k HUIs are computed. Next, a minimum utility threshold θ is fixed dynamically by using the utility of the kth most preferred itemset
w.r.t. utility and Φ is updated according to θ. Then, each time a new itemset is
discovered, the bound θ is fixed accordingly, and so on.

Next, we use a decomposition paradigm to solve the SAT-based Top-k HUIs
enumeration problem. Our aim is to avoid encoding the entire database and to
identify independent sub-problems of reasonable size that will be solved sequentially. Precisely, the main idea is that solving a formula Φ can be splitted into

---

Algorithm 1: DPLL_Enum: A procedure for model enumeration

Input: Φ: a CNF propositional formula
Output: S: the set of models of Φ

1 $\Delta = \emptyset$, $\theta = 0$, $S = \emptyset$
2 **if** ($\Phi \models_* p$) **then**
3 | **return** DPLL_Enum($\Phi \wedge p, \Delta \cup \{p\})$) ;   /* unit clause */
4 **end**
5 **if** ($\Phi \models_* \bot$) **then**
6 | **return** $False$;   /* conflict */
7 **end**
8 **if** ($\Delta \models \Phi$) **then**
9 | $S \leftarrow S \cup \{\Delta\}$;   /* new found model */
10 | **if** ($k \leq |\Delta|$) **then**
11 | | sort(Δ);
12 | | $\theta \leftarrow utility(\Delta[k-1])$;
13 | | update(Φ)
14 | **end**
15 | **return** $False$
16 **end**
17 $p = $ select_variable($Var(\Phi)$);
18 **return** $DPLL\_Enum(\Phi \wedge p, \Delta \cup \{p\}) \vee DPLL\_Enum(\Phi \wedge \neg p, \Delta \cup \{\neg p\})$;
19 **return** S, θ;

---

the enumeration of models of $\Phi \wedge \Psi_1, \ldots, \Phi \wedge \Psi_n$, where $\{\Psi_1, \ldots, \Psi_n\}$ is a set of formulas defined over $Var(\Phi)$ and satisfying the following properties:

(i), $\Psi_i \wedge \Psi_j \models \bot$, $\forall i \neq j$, and (ii) $\Psi_1 \vee \ldots \vee \Psi_n \equiv \top$.

The formulas $\Psi_1, \ldots, \Psi_n$ are defined such that $\Psi_i = p_{a_i} \wedge \bigwedge_{j<i} \neg p_{a_j}$ ($1 \leq i \leq n$). Clearly, Ψ_i requires that the literal p_{a_i} to be assigned $true$, i.e., the itemset should contain a_i while the items a_j with $j < i$ are missing (p_{a_j} is false $\forall j < i$). Such problem can be found by encoding only transactions containing a_i.

Algorithm 2 describes our decomposition technique to compute the set of Top-k HUIs. During the decomposition, an item a is picked, and all transactions that contain a are selected. All items b with $TWU(b, D_{a_i}) < \theta$ are excluded from such transactions. In the Top-k HUIM, θ is fixed to zero until k itemsets are detected. Because θ is fixed uniquely when k HUIs are discovered, we must begin by solving simpler problems until we find a first interesting bound θ value.

4   Experimental Results

We conducted an experimental evaluation using six real-world databases used in the HUIM task: *Chess, Mushroom, Connect, Accidents, Chainstore* and *Retail* [1]. Our Experiments were carried out on an Intel Xeon quad-core processor machine, 32 GB of RAM, and a clock speed of 2.66 GHz. The timeout is fixed to 2 h. We compared our SATTHUIM algorithm to two specialized methods TKO [6], and TKU [4]. We used the open-source data mining library SPMF (Sequential

---

Algorithm 2: SAT based **Top-k** High Utility Itemset Mining (**SATTHUIM**)

Input: A transaction database (D)
Output: The set of all Top-k high-utility itemsets (S)

1 $S \leftarrow \emptyset$;
2 **for** i in $n..1$ **do**
3 | **if** $\underline{TWU(a_i, D_{a_i}) \geq \theta}$ **then**
4 | | $D_{a_i} \leftarrow \{(k, T_k) \in D \mid a_i \in T_k\}$;
5 | | $\Omega = \langle a_1, \ldots, a_n \rangle \leftarrow items(D_{a_i})$;
6 | | $\Gamma \leftarrow \emptyset$;
7 | | **for** $b \in items(D_{a_i})$ **do**
8 | | | **if** $\underline{TWU(b, D_{a_i}) < \theta}$ **then**
9 | | | | $\Gamma \leftarrow \Gamma \cup \{b\}$;
10 | | | **end**
11 | | **end**
12 | | $\Phi \leftarrow encode\_huim\_cnf(D_{a_i}, \theta) \wedge p_{a_i} \wedge \bigwedge\limits_{1 \leq j < i} \neg p_{a_j} \wedge \bigwedge\limits_{b \in \Gamma} \neg p_b$;
13 | | $S \leftarrow S \cup DPLL\_Enum(\Phi, \theta)$;
14 | **end**
15 **end**
16 **return** S;

Pattern Mining Framework) [1]. We also extend our approach to mine Top-k closed HUIs namely SATTCHUIM. We used the MiniSAT [3] as an oracle SAT.

Figure 2 summarizes the empirical performances of our method compared to baselines. In all cases, our method scales than the baselines to all datasets. SATTHUIM achieves interesting results since it performs well in almost of databases when k varies. Second, for some k values, all methods failed to carry out mining tasks. More specifically, TKU fails on *Chess* dataset for $k > 10$, while TKO fails on *Connect* for $k > 10$. On the other hand, our approach fails on *Retail* for $k > 10000$. For *Retail*, the performance of TKO algorithm was an order of magnitude slower compared to our proposed SATTHUIM algorithm. From scalability point of view, our method is able to scale for the different k values under the time limit for large datasets, while TKO and TKU algorithms are not able to scale up on *Retail*, and *Accidents* under the time limit. Lastly, our both approaches require nearly the same time to find all Top-k models, except on *Retail* dataset where SATTHUIM is clearly faster than SATTCHUIM and *Chainstore* dataset where SATTCHUIM is not able to scale.

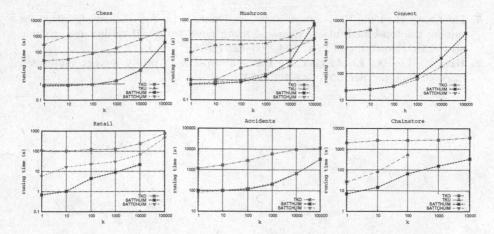

Fig. 2. SATTHUIM and SATTCHUIM vs. baselines on different dataset.

5   Conclusion

In this paper, we tackled the Top-k (closed) high utility itemsets mining problem using propositional satisfiability and a number of well-known established techniques for SAT-based problem solving. Technically, the main idea was to model the Top-k (closed) high utility itemset mining task as a CNF formula such that each of its models corresponds to a Top-k (closed) high utility itemset of interest. Thanks to the decomposition principle, our technique allows to efficiently enumerate the set of Top-k models of the encoding. Despite our promising obtained results, we plan to develop a parallel version to improve the scalability of our SAT-based approach for computing Top-k HUIs for large real-world databases.

References

1. Fournier-Viger, P.: SPMF: A Java Open-Source Data Mining Library. www. philippe-fournier-viger.com/spmf/
2. Hidouri, A., Jabbour, S., Raddaoui, B., Yaghlane, B.B.: A SAT-based approach for mining high utility itemsets from transaction databases. In: Song, M., Song, I.-Y., Kotsis, G., Tjoa, A.M., Khalil, I. (eds.) DaWaK 2020. LNCS, vol. 12393, pp. 91–106. Springer, Cham (2020). https://doi.org/10.1007/978-3-030-59065-9_8
3. Eén, N., Sörensson, N.: An extensible SAT-solver. In: Giunchiglia, E., Tacchella, A. (eds.) SAT 2003. LNCS, vol. 2919, pp. 502–518. Springer, Heidelberg (2004). https://doi.org/10.1007/978-3-540-24605-3_37
4. Wu, C.W., Shie, B.E., Tseng, V.S., Yu, P.S.: Mining top-k high utility itemsets. In: KDD (2012)
5. Ryang, H., Yun, U.: Top k high utility pattern mining with effective threshold raising strategies. Knowl. Based Syst. **76**, 109–126 (2015)
6. Tseng, V.S., Wu, C., Fournier-Viger, P., Yu, P.S.: Efficient algorithms for mining top-k high utility itemsets. IEEE Trans. Knowl. Data Eng. **28**(1), 54–67 (2012)

7. Duong, Q.-H., Liao, B., Fournier-Viger, P., Dam, T.-L.: An efficient algorithm for mining the top-k high utility itemsets, using novel threshold raising and pruning strategies. Knowl. Based Syst. **104**, 106–122 (2016)
8. Han, X., Liu, X., Li, J., Gao, H.: Efficient top-k high utility itemset mining on massive data. Inf. Sci. **557**, 382–406 (2021)

Multi-label Feature Selection Algorithm via Maximizing Label Correlation-Aware Relevance and Minimizing Redundance with Mutation Binary Particle Swarm Optimization

Xiaolin Zhu, Yuanyuan Tao, Jun Li, and Jianhua Xu[✉]

School of Computer and Electronic Information, School of Artificial Intelligence,
Nanjing Normal University, Nanjing 210023, Jiangsu, China
182202039@stu.njnu.edu.cn, {lijuncst,xujianhua}@njnu.edu.cn

Abstract. Multi-label classification deals with a special supervised classification problem where any instance could be associated with multiple class labels simultaneously. As various applications emerge continuously in big data field, their feature dimensionality also increases correspondingly, which generally increases computational burdens and even deteriorates classification performance. To this end, feature selection has become a necessary pre-processing step, in which it is still challenging to design an effective feature selection criterion and its corresponding optimization strategy. In this paper, a novel feature selection criterion is constructed via maximizing label correlation-aware relevance between features and labels, and minimizing redundance among features. Then this criterion is optimized using binary particle swarm optimization with mutation operation, to search for a globally optimal feature selection solution. The experiments on four benchmark data sets illustrate that our proposed feature selection algorithm is superior to three state-of-the-art methods according to accuracy and F1 performance evaluation metrics.

Keywords: Multi-label classification · High-dimensional features · Symmetrical uncertainty · Particle swarm optimization · Mutation operation

1 Introduction

Multi-label classification is a particular supervised classification task in which an instance could belongs to multiple class labels simultaneously and the classes are not exclusive one another [4]. Its application domains cover text categorization, music emotion classification, image annotation and bioinformatics.

Supported by Natural Science Foundation of China (NSFC) under Grants 62076134 and 61703096.

As many applications emerge continuously, besides large instance size, their feature dimensionality also goes high and high, which unavoidably includes some possible redundant and irrelevant features. This situation usually increases computational costs, and even deteriorates classification performance. To cope with this difficulty, feature selection (FS) originally used in single-label paradigm is also applied to multi-label case, to select some most discriminative features [6,15]. Therefore, multi-label FS task has become a hot issue in big data [6,15].

As in single-label case, existing multi-label FS techniques are categorized into three groups: embedded, wrapper and filter methods, according to whether a specific classifier is directly involved in FS stage [6,15]. Embedded methods execute a complicated optimization process to both realize FS selection and classifier design. Wrapper methods detect some discriminative features according to the classification performance of some fixed classifier. Filter methods evaluate the quality of features (or feature subset) on the basis of the intrinsic characteristics and structures of data, without using any classifier. Generally, the last group of FS methods are more efficient than, and as effective as the first two ones, therefore filter methods become a hot research point in multi-label FS field.

For filter FS methods, there are two basic elements: a proper feature (or feature subset) evaluation criterion and a related optimization technique, which is associated with the effectiveness and efficiency of FS methods greatly. As we known, mutual information (MI) and its variants are some widely-used information theory based FS criteria, resulting in three kinds of effective and efficient FS methods via simple ranking, greedy search and evolutionary computation optimization strategies [15,18].

Ranking based methods evaluate relevances between features and labels, then sort them in descending order, and finally select some top ranked features as an FS solution. Pruning problem transformation (PPT) is a special data decomposition that deletes some label combinations with a few instances. Based on PPT, PPT-MI is introduced as a byproduct in [8], which directly applies MI to evaluate relevance. To depict label correlations, Lee and Kim [9] further estimate relevance between one feature and two labels based on three variable MI, which results in an efficient ranking-based FS method (FIMF). However, these methods do not characterize redundance among features explicitly.

Greedy search based methods search for a discriminative feature subset step by step, where sequential forward selection (SFS) is a widely-used optimization strategy. Besides the first-order relevance between one feature and one label, PMU [8] takes second-order interaction in account. This criterion is extended to high-order interaction between features and labels in MAMFS [10]. Both maximizing relevance and minimizing redundancy (mRMR) via SFS strategy are one of most cited single-label FS methods [14], which is generalized to deal with multi-label FS task using different ways. In MDMR [12], a three variable MI term is added to its redundance. In GFS-ML [11], the converted label granules substitute for original labels for estimating relevance. These SFS-based multi-label FS methods theoretically only find out a locally optimal feature subset.

Evolutionary computation based methods apply evolutional algorithms (e.g., genetic algorithm (GA)) and swarm intelligence (e.g., particle swarm optimization (PSO), ant colony optimization (ACO), and bat algorithm (BA)), to find out a globally optimal FS subset [13]. In [2], MI is normalized into symmetrical uncertainty for single-label FS case. After this criterion is extended to multi-label case using problem decomposition, both GA and BA are applied in [5,17], respectively. Besides maximizing relevance in [8], minimizing redundance among features is added in [19], which is optimized by binary PSO. A similar mRMR criterion for multi-label FS case is optimized by ACO [3]. Due to possible global search ability, improving this kind of optimization techniques and designing a novel FS criterion are still an open problem for multi-label FS task.

In this paper, we apply symmetrical uncertainty [2] to depict relevance between features and labels and redundance among features, and Jaccard coefficient to characterize label correlation, to construct a new FS criterion (simply SUC) to maximize label correlation-aware relevance and to minimize redundance. Then a modified particle swarm optimization (i.e., MBPSO) is built, which includes a round and modular operation to binarize continuous positions and a mutation operation to adjust the number of selected features. Finally a novel multi-label FS selection is proposed via combining SUC with MBPSO, which is named as SUC-PSO concisely. Our experiments from four benchmark data sets show that our proposed FS method performs better than three existing approaches [8,9,19] according to two instance-based performance metrics: accuracy and F1.

2 A Novel Multi-label Feature Selection Method

2.1 Preliminaries

There is a given multi-label training data set which consists of a real feature matrix $\mathbf{X} \in \mathcal{R}^{D \times N}$ and a binary label matrix $\mathbf{Y} \in \{0,1\}^{C \times N}$, i.e.,

$$\mathbf{X} = [\mathbf{x}_1, ..., \mathbf{x}_i, ..., \mathbf{x}_N], \mathbf{Y} = [\mathbf{y}_1, ..., \mathbf{y}_i, ..., \mathbf{y}_N] \qquad (1)$$

where N, D and C are the numbers of instances, features and labels, respectively. The i-th instance is represented by its column feature vector $\mathbf{x}_i = [x_{i1}, ..., x_{iD}]^T$ $\in \mathcal{R}^D$ and label one $\mathbf{y}_i = [y_{i1}, ..., y_{iC}]^T \in \{0,1\}^C$. The original feature and label index sets are $\mathcal{F} = \{f_1, ..., f_D\}$ and $\mathcal{L} = \{l_1, ..., l_C\}$, respectively. The multi-label feature selection chooses a discriminative feature subset $\mathcal{S}$ of size d from $\mathcal{F}$ ($d < D$) to remain those highly relevant and lowly redundant features.

2.2 Feature Selection Criterion Based on Symmetrical Uncertainty

For two random variables A and B, the information entropy $H(A)$ measures the uncertainty of random variable A only, and the mutual information $I(A, B)$

describes the shared information between A and B [18]. To execute feature selection task better, the symmetrical uncertainty [2] is defined as:

$$SU(A, B) = 2I(A, B)/(H(A) + H(B)) \qquad (2)$$

which will be used to describe the relevance between features and labels, and redundance among features, in this study.

In single-label FS situation, a widely-used FS criterion is to minimize redundance among features, and to maximize relevance between features and labels (mRMR) in [14], which is described as follows:

$$\max \frac{1}{|\mathcal{S}|} \sum_{f_i \in \mathcal{S}} \sum_{l_j \in \mathcal{L}} I(f_i, l_j) - \frac{1}{|\mathcal{S}|^2} \sum_{f_i, f_j \in \mathcal{S}} I(f_i, f_j) \qquad (3)$$

where $|\mathcal{S}|$ is the size of subset $\mathcal{S}$. This mRMR is executed by SFS strategy, resulting in a locally optimal FS solution. For our multi-label case, to depict possible correlation among labels, the following correlation label matrix $\boldsymbol{\Theta} = [\theta_{ij} | i, j = 1, ..., C]$ with Jaccard correlation coefficients [16] is adopted

$$\theta_{ij} = \begin{cases} 0, \text{if } i = j \\ \frac{\|\mathbf{y}^i \circ \mathbf{y}^j\|_1}{\|\mathbf{y}^i\|_1 + \|\mathbf{y}^j\|_1 + \|\mathbf{y}^i \circ \mathbf{y}^j\|_1}, \text{otherwise} \end{cases} \qquad (4)$$

where "$\circ$" is the Hadamard product of two vectors and $\|\cdot\|_1$ is the 1-norm of vector. Then the above criterion (3) is extended as

$$\max \sum_{f_k \in \mathcal{S}} \sum_{l_i, l_j \in \mathcal{L}, j \neq i} I(f_k; l_i)\theta(l_i, l_j) + \sum_{f_k \in \mathcal{S}} \sum_{l_i \in \mathcal{L}} I(f_k; l_i) - \sum_{f_i, f_j \in \mathcal{S}} I(f_i; f_j). \qquad (5)$$

Due to $\theta_{ii} = 0$, we introduce a quantity $\bar{\theta}(l_i)(i = 1, ..., C)$:

$$\bar{\theta}(l_i) = \sum_{l_j \in \mathcal{L}} \theta_{ij} \qquad (6)$$

to estimate the correlation of the i-th label over all labels but itself, and then simplify the above criterion (5) into

$$\max \sum_{f_k \in \mathcal{S}} \sum_{l_i \in \mathcal{L}} I(f_k; l_i)(1 + \bar{\theta}(l_i)) - \sum_{f_i, f_j \in \mathcal{S}} I(f_i; f_j). \qquad (7)$$

Further, we substitute SU (2) for MI to define a new multi-label FS criterion:

$$\max \sum_{f_k \in \mathcal{S}} \sum_{l_i \in \mathcal{L}} SU(f_k, l_i)(1 + \bar{\theta}(l_i)) - \sum_{f_i, f_j \in \mathcal{S}} SU(f_i, f_j). \qquad (8)$$

This criterion (simply SUC) is to minimize redundance among features and maximize relevance between features and labels, under consideration of label correlation information, according to symmetrical uncertainty.

Table 1. Four multi-label benchmark data sets used in our experiments.

Dataset	#Domain	#Training	Testing	#Features	#Classes	#Average labels
Emotions	Music	391	202	72	6	1.87
Image	Image	1200	800	294	5	1.24
Virus	Biology	124	83	440	6	1.22
Yeast	Biology	1500	917	103	14	4.24

Table 2. Comparison of four algorithms on **Emotions** data set

Method	Metric	10%	20%	30%	40%	50%	60%	70%	80%	90%	100%
FIMF	AC(↑)	**42.61**	45.66	44.52	40.98	48.07	48.44	48.72	49.05	55.56*	50.08
	F1(↑)	**50.67**	52.92	52.93	48.97	55.59	55.92	55.80	58.20	64.43*	58.46
PMU	AC(↑)	31.85	48.18	47.24	47.07	47.38	43.61	47.81	47.54	51.11*	50.08
	F1(↑)	38.05	55.86	54.77	54.50	54.54	51.57	56.25	55.02	59.59*	58.46
MMI-PSO	AC(↑)	22.93	36.38	43.11	43.06	47.77	50.53*	49.50*	54.12*	53.01*	50.08
	F1(↑)	28.59	43.56	51.93	51.60	56.58	58.89*	57.62*	**62.82***	**61.30***	58.46
SUC-PSO	AC(↑)	26.56	**47.97**	**47.97**	**52.18***	**50.78***	**52.51***	**51.85***	**54.92***	**57.74***	50.08
	F1(↑)	32.62	**55.49**	**56.79**	**59.93***	**58.54***	**60.46***	**59.67***	61.94*	58.20*	58.46

2.3 Multi-label Feature Selection Algorithm with Binary Particle Swarm Optimization with Mutation Operation

To optimize the aforementioned FS criterion (8), we apply and modify particle swarm optimization (PSO) to search for a globally optimal feature subset.

The original PSO is a population-based stochastic method for solving continuous optimization problems. In such a PSO, its each individual is called as a particle, which moves in the search space of a given optimization problem. The position z_i of the i-th particle represents a possible solution. Each particle tries to find out its better positions in the search space by changing its velocity according to some rules inspired by behavioral models of bird flocking [1,13].

For our feature selection task, the previous continuous solutions need to be reduced into a binary one, where 1 or 0 means whether a feature is selected or not. Therefore binary PSO (simply BPSO) was proposed [13], via the sigmoid function and a proper threshold. In this study, we adopt a new binarization way originally for artificial bee colony algorithm [7], which applies round operator and double modular process:

$$z_{ij} = \text{round}(|z_{ij}| \bmod 2) \bmod 2 \tag{9}$$

where z_{ij} denotes the j-th component of z_i, which is added into the above original PSO to construct BPSO in this paper.

The aforementioned BPSO could find out a solution with an optimal fitness function value. However, due to random optimization strategy, the number of "1" components of selected features is not fixed during its executing procedure. In this paper, to choose a fixed size for feature subset, we further add a mutation operation from GA to adjust the number of "1" components to be a fixed value

Table 3. Comparison of four algorithms on **Image** data set

Method	Metric	10%	20%	30%	40%	50%	60%	70%	80%	90%	100%
FIMF	AC(↑)	19.58	24.00	31.18	33.22	33.91	38.56	43.12	45.72	47.35	47.54
	F1(↑	21.20	25.43	33.02	34.87	35.75	40.83	45.47	48.27	49.75	50.12
PMU	AC(↑)	33.08	35.65	33.98	40.58	41.88	43.71	41.02	45.08	46.38	47.54
	F1(↑)	34.92	37.67	35.65	42.73	44.25	46.06	43.02	47.35	48.83	50.12
MMI-PSO	AC(↑)	**40.16**	41.29	48.97	46.39	45.72	48.33	45.87	42.22	47.04	47.54
	F1(↑	**42.33**	44.00	50.93	49.02	48.08	50.08	48.33	44.62	49.77	50.12
SUC-PSO	AC(↑)	34.85	**46.31**	**46.60**	**42.85**	**50.54***	**45.04**	**50.58***	**46.66**	**49.71***	47.54
	F1(↑)	36.91	**49.45**	**49.45**	**45.29**	**53.33***	**47.58**	**53.60***	**49.22**	**52.69***	50.12

Table 4. Comparison of four algorithms on **Yeast** data set

Method	Metric	10%	20%	30%	40%	50%	60%	70%	80%	90%	100%
FIMF	AC(↑)	40.28	42.19	44.47	44.80	46.04	46.63	45.31	46.96	50.01*	49.20
	F1(↑)	51.92	53.42	55.11	55.50	56.59	57.37	55.89	57.70	60.59*	59.92
PMU	AC(↑)	44.39	44.29	43.48	44.80	46.04	46.63	45.31	46.96	50.01*	49.20
	F1(↑)	55.66	55.04	54.19	55.50	56.59	57.37	55.89	57.70	60.59*	59.92
MMI-PSO	AC(↑)	40.50	43.49	46.80	48.73	46.71	48.15	**49.65***	49.02	49.14	49.20
	F1(↑)	51.88	54.33	58.11	59.75	57.36	59.12	**60.71***	59.86	59.96*	59.92
SUC-PSO	AC(↑)	**46.11**	**47.18**	**45.48**	**49.86***	**49.86***	**49.28***	49.28*	**50.19***	**50.23***	49.20
	F1(↑)	**57.20**	**58.40**	**56.61**	**60.61***	**60.56***	**60.15***	59.93*	**60.83***	**60.80***	59.92

d, through removing and adding "1" elements correspondingly. We referred to our above stated PSO version as MBPSO simply in this paper.

Finally, we apply our MBPSO to optimize our FS criterion SUC (8) to construct a novel multi-label FS approach (simply SUC-PSO).

3   Experiments and Analysis

In this section, we evaluate our FS method SUC-PSO using four multi-label benchmark data sets via comparing with three existing FS methods (FIMF [9], PMU [8] and MMI-PSO [19]).

3.1   Four Benchmark Data Sets and Two Evaluation Metrics

In this paper, we downloaded four widely-validated benchmark data sets: Emotions, Image, Virus and Yeast form[1], to evaluate and compare our algorithm and other existing feature selection methods, as shown in Table 1. This table also provides some important statistics for these sets.

Further, we utilize two multi-label classification performance metrics (accuracy (AC) and F1) from multi-label k nearest neighbor method (ML-kNN) [20] with its recommended setting ($k = 10$ and the smooth factor is 1), to evaluate and compare the performance of different FS methods. On the definitions of two metrics, please refer to [4]. The higher such two metrics are, the better the FS methods work, as shown in upper arrow (↑) in our experimental tables.

---

[1] http://ceai.njnu.edu.cn/Lab/LABIC/LABIC_Software.html.

Table 5. Comparison of four algorithms on **Virus** data set

Method	Metric	10%	20%	30%	40%	50%	60%	70%	80%	90%	100%
FIMF	AC($\uparrow$)	43.36	45.62	48.41	48.99	50.77	52.20	49.90	52.07	57.09*	54.91
	F1($\uparrow$)	41.95	43.40	42.95	43.08	43.21	43.53	43.97	44.41	45.61	45.84
PMU	AC($\uparrow$)	49.46	48.83	46.71	48.99	50.77	52.20	49.90	52.07	57.09*	54.91
	F1($\uparrow$)	42.61	42.93	43.45	43.08	43.21	43.53	43.97	44.41	45.61	45.84
MMI-PSO	AC($\uparrow$)	44.18	47.97	**52.73**	56.13	51.61	54.58	**57.07***	55.92*	54.98*	54.91
	F1($\uparrow$)	42.37	42.49	**44.45**	44.31	44.21	45.32	45.82	45.60	45.70	45.84
SUC-PSO	AC($\uparrow$)	**51.87**	**54.57**	51.19	**57.42***	**56.49***	**55.83***	55.63*	**56.58***	**57.16***	54.91
	F1($\uparrow$)	**43.80**	**43.99**	43.33	**44.68**	**46.05***	**45.48**	**46.16***	45.75	**45.86***	45.84

Table 6. Comparison of "top" and "win" for four compared FS techniques

Index	FIMF	PMU	MMI-PSO	SUC-PSO
Top	4	0	9	**61**
Win	3	5	14	**39**

3.2 Experimental Results and Analysis

In this sub-section, we report our experimental results from four methods and four data sets. To evaluate the classification performance of each method comprehensively, we regard two metrics as functions of the proposition of selected features from 10% to 90% with a step 10%, as shown in Tables 2, 3, 4, 5. We show two special annotations for these tables. One is the **bold font** to indicate which method perform the best across four method at the specific feature proposition. The other is the red star (*) to indicate that a specific method achieves better value at some proposition, compared with the last column value with all features.

From these four tables, it is observed that at most of propositions of selected features, our SUC-PSO performs the best, compared with three existing methods. In order to compare these four multi-label FS methods extensively, we use two indexes: top and win, where the top indicates how many times some FS method performs the best among four methods, i.e., the number of metric values in bold font, and the win represents how many times some FS method is superior to ML-kNN without FS, i.e., the number of metric values with red star. These two indexes are shown in Table 6. It is found out that SUC-PSO obtains 61 tops and 39 wins, respectively, which are much higher than the sum from other three FS methods.

Overall, from the above experimental results and analysis, it could be illustrated that our proposed FS method is more effective, compared with three state-of-the-art methods (FIMF, PMU and MMI-PSO).

4 Conclusions

In this paper, on the one hand, we design a new multi-label feature selection criterion to maximize label correlation-aware relevance between features and labels, and minimize redundance among features, according to symmetrical uncertainty. On the other hand, we propose a special particle swarm optimization version via round and modular operation based binarization and mutation operation to search for a fixed size feature subset. Integrating such two techniques is to construct a novel multi-label feature selection approach. Our extensive experimental results on four benchmark data sets show the effectiveness of our proposed FS method, compared with three state-of-the-art existing FS approaches, according to two performance evaluation metrics (accuracy and F1). For future work, we will execute more experiments on more data sets and compare more existing FS techniques to validate classification performance of our method in this paper.

References

1. Freitas, D., Lopes, L.G., Morgado-Dias, F.: Particle swarm optimisation: a historical review up to the current developments. Entropy **22**, Article-362 (2020)
2. Hall, M.A.: Correlation-based feature selection for discrete and numeric class machine learning. In: 17th International Conference on Machine Learning (ICML 2000), pp. 359–366. OmniPress, Madson WI, USA (2000)
3. Hatami, M., Mehrmohammadi, P., Moradi, P.: A multi-label feature selection based on mutual information and ant colony optimization. In: 28th Iranian Conference Electrical Engineering (ICEE 2020), pp. 1–6. IEEE Press, New York, USA (2020)
4. Herrera, F., Charte, F., Rivera, A.J., del Jesus, M.J.: Multilabel Classification Problem Analysis. Metrics and Techniques. Springer, Cham (2016). https://doi.org/10.1007/978-3-319-41111-8
5. Jungjit, S., Freitas, A.A.: A new genetic algorithm for multi-label correlation-based feature selection. In: 23rd European Symposium Artificial Neural Network, Artificial Intelligence Machine Learning (ESANN 2015), pp. 285–290. CIACO Press, Belgium (2015)
6. Kashef, S., Nezamabadi-pour, H., Nipour, B.: Multilabel feature selection: a comprehensive review and guide experiments. WIREs Data Min. Knowl. Disc, **8**(2), Article-e1240 (2018)
7. Kiran, M.S.: The continuous artificial bee colony algorithm for binary optimization. Appl. Soft Comput. **33**, 15–23 (2015)
8. Lee, J., Kim, D.W.: Feature selection for multi-label classification using multivariate mutual information. Pattern Recogn. Lett. **34**(3), 349–357 (2013)
9. Lee, J., Kim, D.W.: Fast multi-label feature selection based on information-theoretic feature ranking. Pattern Recogn. **48**(9), 2761–2771 (2015)
10. Lee, J., Kim, D.W.: Mutual information-based multi-label feature selection using interaction information. Expert Syst. Appl. **42**(4), 2013–2025 (2015)
11. Li, F., Miao, D., Pedrycz, W.: Granular multi-label feature selection based on mutual information. Pattern Recogn. **67**, 410–423 (2017)
12. Lin, Y., Hu, Q., Liu, J., Duan, J.: Multi-label feature selection based on max-dependency and min-redundancy. Neurocomputing **168**, 92–103 (2015)
13. Nguyen, B.H., Xue, B., Zhang, M.: A survey on swarm intelligence approaches to feature selection in data mining. Swarm Evol. Comput. **54**, Article-100663 (2020)
14. Peng, H., Long, F., Ding, C.: Feature selection based on mutual information: criteria of max-dependency, max-relevance, and min-redundancy. IEEE Trans. Pattern Anal. Mach. Intell. **27**(8), 1226–1238 (2005)

15. Siblini, W., Kuntz, P., Meyer, F.: A review on dimensionality reduction for multi-label classification. IEEE Trans. Knowl. Data Eng. **33**(3), 839–857 (2021)
16. Sun, Z., et al.: Mutual information based multi-label feature selection via constrained convex optimization. Neurocomputing **329**, 447–456 (2019)
17. Tao, Y., Li, J., Xu, J.: Multi-label feature selection method via maximizing correlation-based criterion with mutation binary bat algorithm. In: 32nd International Joint Conference Neural Networks (IJCNN 2020), pp. 1–8. IEEE Press, New York, USA (2020)
18. Vergara, J.R., Estévez, P.A.: A review of feature selection methods based on mutual information. Neural Comput. Appl. **24**(1), 175–186 (2013). https://doi.org/10.1007/s00521-013-1368-0
19. Wang, X., Zhao, L., Xu, J.: Multi-label feature selection method based on multivariate mutual information and particle swarm optimization. In: Cheng, L., Leung, A.C.S., Ozawa, S. (eds.) ICONIP 2018. LNCS, vol. 11304, pp. 84–95. Springer, Cham (2018). https://doi.org/10.1007/978-3-030-04212-7_8
20. Zhang, M.L., Zhou, Z.H.: ML-KNN: a lazy learning approach to multi-label learning. Pattern Recogn. **40**(7), 2038–2048 (2007)

Mining Partially-Ordered Episode Rules
with the Head Support

Yangming Chen[1], Philippe Fournier-Viger[1(✉)] (iD), Farid Nouioua[2],
and Youxi Wu[3]

[1] Harbin Institute of Technology (Shenzhen), Shenzhen, China
[2] University of Bordj Bou Arreridj, El-Anasser, Algeria
[3] Hebei University of Technology, Tianjin, China

Abstract. Episode rule mining is a popular data analysis task that aims
at finding rules describing strong relationships between events (or symbols) in a sequence. Finding episode rules can help understanding the
data or making predictions. However, traditional episode rule mining
algorithms find rules that require a very strict ordering between events.
To loosen this ordering constraints and find more general and flexible
rules, this paper presents an algorithm named POERMH (Partially-
Ordered Episode Rule Mining with Head Support). Unlike previous algorithms, the head support frequency measure is used to select interesting episode rules. Experiments on real data show that POERMH can
find interesting rules that also provides a good accuracy for sequence
prediction.

Keywords: Episode rules · Partially ordered rules · Head support

1 Introduction

Discrete sequences of events or symbols is a data type found in many domains.
For instance, a sequence may represents words in a text, nucleotides in a DNA
sequence, and locations visited by a tourist in a city. To analyze a sequence
of symbols, a popular data science task is episode mining [1,5–7]. The aim is
to find subsequences that appear many times, called *episodes*, and rules called
episode rules. These rules indicate strong sequential relationships between events
or symbols in a sequence, and can help humans to understand the data take
decisions, and also be used to predict the next symbol (event) of a sequence [1,
3,6,7].

Though several episode rule mining algorithms were proposed, a problem
is that they enforce a very strict ordering of events [3]. For instance, a rule
$\langle e, f \rangle \rightarrow \langle a \rangle$ indicates that if symbol e is followed by f, then it will be followed
by a. Though this rule is easily understandable, the ordering between events is
very strict and thus a similar rule with a different ordering such as $\langle f, e \rangle \rightarrow \langle a \rangle$
is viewed as different. As a solution to this issue, an algorithm named POERM
was proposed to discover a more general type of rules, called *partially-ordered*

© Springer Nature Switzerland AG 2021
M. Golfarelli et al. (Eds.): DaWaK 2021, LNCS 12925, pp. 266–271, 2021.
https://doi.org/10.1007/978-3-030-86534-4_26

episode rules (POER) [3]. A POER has the form $X \rightarrow Y$, indicating that if some symbols X appear in any order, they will be followed by some symbols Y.

Even though POERM was shown to find interesting rules in real data, a drawback is that each rule is evaluated by only counting its non-overlapping occurrences. But it was argued in other data mining studies that counting occurrences of episodes based on a concept of sliding windows has several advantages over counting non-overlapping occurrences [6]. For this reason, this paper introduces a modified version of POERM, called POERMH (POERM with Head support), that relies on the *head support* measure [6] to find episode rules.

The rest of this paper is divided into four sections. Section 2 describes preliminaries and the problem definition. Section 3 presents the designed POERMH algorithm. Section 4 reports results of experiments. Finally, Sect. 5 presents a conclusion and discusses future work.

2   Problem Definition

Episode mining aims at finding interesting patterns in a sequence of events, annotated with timestamps. Let $E = \{e_1, e_2, \dots e_n\}$ be a finite set of event types. A set of distinct events $R \subseteq E$ is called an **event set**. An **event pair** is a tuple $P = (t, R)$ indicating that a non empty set of events $R \subseteq E$ occurred at some time t. An **event sequence** $S = \langle (t_1, R_1), (t_2, R_2), \dots, (t_m, R_m) \rangle$ is a time-ordered list of event pairs, where $R_i \subseteq E$ for $1 \leq i \leq m$ and $0 \leq t_1 < t_2 < \dots < t_m$. For instance, consider the following event sequence: $S = \langle (1, \{c\}), (2, \{a, b\}), (3, \{d\}), (5, \{a\}), (6, \{c\}), (7, \{b\}), (8, \{d\}), (10, \{a, b, c\}), (11, \{a\}) \rangle$. This sequence contains nine event sets having timestamps ranging from 1 to 11, and five event types $E = \{a, b, c, d, e\}$. The sequence means that event c appeared at time 1, was followed by events a and b at time 2, then by d at time 3, and so on.

In recent work, it was proposed to discover a novel type of episode rules called partially-ordered episode rules (POERs) [3] to loosen the strict ordering constraint of traditional episode rules. A **partially-ordered episode rule** has the form $Y_1 \rightarrow Y_2$, where Y_1 and Y_2 are two event sets such that $Y_1 \neq \emptyset$ and $Y_2 \neq \emptyset$. Such rule is interpreted as if the events in Y_1 are observed, they are followed by those of Y_2.

Two evaluation functions have been proposed to find interesting POERs in a sequence of events, which are the non-overlapping support and confidence. They are defined as follows, based on the concept of occurrence.

Definition 1 (Occurrence). *Let there be a sequence* $S = \langle (t_1, R_1), (t_2, R_2), \dots, (t_v, R_v) \rangle$ *and three user-specified time constraints* $\eta, winlen, \alpha \in \mathbb{Z}^+$. *Moreover, let there be an event set* Y *and a POER* $Y_1 \rightarrow Y_2$. *The set* Y **has an occurrence** *in* S *for the time interval* $[t_i, t_j]$ *if* $Y \subseteq R_i \cup R_{i+1} \dots \cup R_j$. *The rule* $Y_1 \rightarrow Y_2$ **has an occurrence** *in* S *for the time interval* $[t_i, t_j]$ *if some timestamps* t_v *and* t_w *exist such that* Y_1 *has an occurrence in* $[t_i, t_v]$, Y_2 *has an occurrence in* $[t_w, t_j]$, $t_i \leq t_v < t_w \leq t_j$, $t_j - t_i < winlen$, $t_j - t_w < \eta$, *and also* $t_v - t_i < \alpha$.

Definition 2 (Non-overlapping support and confidence.). *Let there be a sequence S, an event set Y, and a rule $Y_1 \to Y_2$. The set of occurrences of Y in S is denoted as $occ(Y)$. And the set of occurrences of $Y_1 \to Y_2$ in S is denoted as $occ(Y_1 \to Y_2)$. An occurrence $[t_{i1}, t_{j1}]$ is called **redundant** in a set of occurrences if there is an overlapping occurrence $[t_{i2}, t_{j2}]$ where $t_{i1} \leq t_{i2} \leq t_{j1}$ or $t_{i2} \leq t_{i1} \leq t_{l2}$. Let $nocc(Y)$ denotes the **set of all non redundant occurrences** of Y. Moreover, let $nocc(Y_1 \to Y_2)$ denotes the **set of all non redundant occurrences of** $Y_1 \to Y_2$ in S. The **non-overlapping support** of $Y_1 \to Y_2$ is defined as $sup(Y_1 \to Y_2) = |nocc(Y_1 \to Y_2)|$. The **non-overlapping confidence** of $Y_1 \to Y_2$ is defined as $conf(Y_1 \to Y_2) = |nocc(Y_1 \to Y_2, S)|/|nocc(Y_1)|$.*

Then, the goal of episode rule mining is to find all episode rules having a support and confidence that is no less than some minimum thresholds $minsup$ and $minconf$ [3]. Though this definition can be useful, it was argued in other papers that considering a window based definition of the support may provide more interesting patterns [6]. Hence, this paper defines two novel measures to replace the non-overlapping support and confidence.

Definition 3 (Head support). *Let there be a POER $Y_1 \to Y_2$. The **head support of** $Y_1 \to Y_2$ in a sequence S is defined as $sup(Y_1 \to Y_2, S) = |wocc(Y_1 \to Y_2)|$ where $wocc(Y_1 \to Y_2)$ is the number of occurrences of the rule that have distinct start points.*

Definition 4 (Head confidence). *Let there be a POER $Y_1 \to Y_2$. The **non-overlapping confidence** of $Y_1 \to Y_2$ is defined as $conf(Y_1 \to Y_2) = |wocc(Y_1 \to Y_2, S)|/|wocc(Y_1)|$, where $wocc(Y_1)$ is the number of occurrences of the rule that have distinct start points.*

Then, the problem studied in this paper is to find all episode rules having a head support and confidence that are not less than some minimum thresholds $minsup$ and $minconf$ [3].

For instance, Let there be $minsup = 3$, $minconf = 0.6$, $\alpha = 3$, $winlen = 5$, $\eta = 1$ and the sequence: $S = \langle (1, \{c\}), (2, \{a,b\}), (3, \{d\}), (6, \{a\}), (7, \{c\}), (8, \{b\}), (9, \{a\}), (11, \{d\}) \rangle$.

The event set $\{a,b,c\}$ has three occurrences in S, which are $wocc(\{a, b, c\}, S) = \{[t_1, t_2], [t_6, t_8], [t_7, t_9]\}$. The start points of these occurrences are t_1, t_6 and t_8, respectively. The rule $r : \{a, b, c\} \to \{d\}$ has two occurrences: $wocc(r, S) = \{[t_1, t_3], [t_7, t_{11}]\}$. Hence, $supp(r, S) = 3$, $conf(r, S) = 2/3$, and R is a POER.

3   The POERMH Algorithm

POERMH receives as input an event sequence and parameters $minsup$, $minconf$, α, $winlen$, and η. The first step of POERMH is to find the frequent rule antecedents. Because POERMH only considers frequent events, the algorithm scans the input sequence to discover frequent 1-event sets and record their position lists. Then, for each frequent 1-event, POERMH scans its position lists, and for each position, POERMH tries to extend it to make 2-event sets. Then, for each i-event set, POERMH counts its head frequency occurrences, record the frequent event sets and extends them into $i+1$-event sets. This process is applied until all frequent event sets are found.

To count the head frequency occurrences, each event sets' position list is read. Firstly, the algorithm applies the quick sort algorithm to positions by ascending start point. Then, a variable *started* is initialized with the default value of -1. This variable contains the most recent starting point that is covered. Then, the sequence is read and each position that has a starting point that is less than *started* is ignored. For any other starting point, it is used to update *started*. At the same time, the number of valid positions is stored in a variable named *support*. After scanning the position list, *support* contains the head frequency occurrences of this event set.

The second step of POERMH is to find all consequents for each antecedent event set to make rules. This process is similar to the POERM algorithm but has some differences. For each position of an antecedent event set, it is expressed in the form of $[pos.start, pos.end]$, and POERMH searches the position $[pos.end+1, winlen - pos.start]$ to find the consequent event sets. Different from POERM, as long as the location of one antecedent's occurrence is fixed, the range of its consequents is also fixed. Hence, a bitmap can be used for each event set to represent whether it appears after a certain position in the antecedent event sets. Then, two event sets can be compared by applying the AND operation with those two bitmaps to get a new bitmap representing the new event sets's head frequency occurrences. Then the support divided by the antecedent's support is calculated to get the confidence. If the confidence is greater than *minconf*, a valid POER is obtained. After searching all the antecedents, the algorithm terminates by returning all the valid rules.

The process of calculating the head frequency of occurrences using a bitmap is as follows. For each sorted position list obtained during the first step, a variable *started_rule* is initialized with the default value of -1. It stores the most recent starting point that is covered. Moreover, a variable *support_rule* is used to record the rule' support. Then the sorted position list is traversed to find a position such that its corresponding position in the bitmap is marked as 1. If the position's starting point is greater than *started_rule*, the *support_rule* variable is increased by one and the *started_rule* variable is updated to the starting point. And finishing scanning the position list, *started_rule* contains the head frequency occurrences of the rule.

The detailed pseudocode of POERMH is not shown due to space limitation but the Java source code is available in the SPMF pattern mining library [4].

4   Experiments

Due to space constraints, the paper only compares the performance of the POERM and POERMH algorithms on two public datasets often used in episode and pattern mining studies. Those datasets can be downloaded from the SPMF pattern mining library [4]. The *FIFA* dataset is a sequence of 710,435 click events on the FIFA World Cup 98 website, where there are 2,990 distinct event types. The *Leviathan* dataset is a sequence of words obtained by transforming the Leviathan novel written by Thomas Hobbes (1651). The sequence has 153,682 words from 9,025 distinct word types.

Each dataset was split into two files: a **training file** (the first 75% of a sequence) and a **testing file** (the remaining 25% of the sequence). The training file is used to discover episode rules, while the testing file is used to assess the quality of predictions using these rules. For this experiment, a window of a length called **winlen** is slide over the test sequence. For each position of that window, given the first **prefix length** items, the goal is to predict the event following the **prefix length** but within **winlen** using the rules. Based on the prediction results, two measures were calculated. The **accuracy** is how many good predictions were done divided by the number of opportunities for prediction (the number of windows in the test set). The **matching rate** is how many good or bad predictions were done, divided by the number of prediction opportunities.

The experiment consisted of varying the prefix size and applying the POERM and POERMH algorithms with a minimum confidence of 30% and 70% to assess the influence of these parameters on the accuracy and matching rate. Figure 1 shows the results obtained by varying the prefix size parameter from 2 to 6 on the two datasets. It is observed that the prefix length has a strong influence on results. Increasing the prefix length generally results in more rules, which increases the matching rate. But at some point increasing the prefix size has less and less influence, which is reasonable. In terms of accuracy, it can decreases when the prefix length is increased but this is due to the increase in matching rate. In terms of confidence, it is found that POERM and POERMH have better accuracy and matching rate when run with a minimum confidence of 30% instead

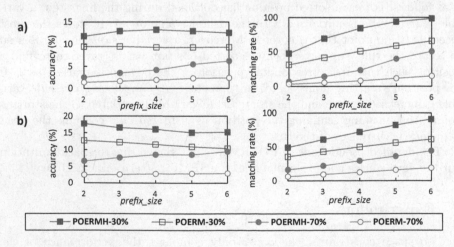

Fig. 1. Influence of *prefix_size* on (a) Bible and (b) Leviathan

of 70%, as more rules are found. On overall, POERMH's accuracy and matching rate are respectively up to 5% and 17% higher than POERM.

Due to space constraints, results about the time efficiency and memory usage are not presented. But in general, the time and space consumed by POERM and POERMH are about the same. POERMH thus provides an improvement over POERM as it can obtain higher accuracy and matching rate while consuming about the same amount of resources.

5   Conclusion

This paper has presented a new algorithm to find partially-ordered episode rules named POERMH. This algorithm relies on the head support measure to identify interesting episode rules. Experiments have shown that the algorithm has good sequence prediction performance compared to the benchmark POERM algorithm. In future work, we will consider adding other optimizations to POERMH, developing a version for distant event prediction [2] and for online mining [9], and finding other types of episode rules [8].

References

1. Ao, X., Luo, P., Wang, J., Zhuang, F., He, Q.: Mining precise-positioning episode rules from event sequences. IEEE Trans. Knowl. Data Eng. **30**(3), 530–543 (2017)
2. Fahed, L., Lenca, P., Haralambous, Y., Lefort, R.: Distant event prediction based on sequential rules. Data Sci. Pattern Recogn. **4**(1), 1–23 (2020)
3. Fournier-Viger, P., Chen, Y., Nouioua, F., Lin, J.C.-W.: Mining partially-ordered episode rules in an event sequence. In: Nguyen, N.T., Chittayasothorn, S., Niyato, D., Trawiński, B. (eds.) ACIIDS 2021. LNCS (LNAI), vol. 12672, pp. 3–15. Springer, Cham (2021). https://doi.org/10.1007/978-3-030-73280-6_1
4. Fournier-Viger, P., et al.: The SPMF open-source data mining library version 2. In: Berendt, B., et al. (eds.) ECML PKDD 2016. LNCS (LNAI), vol. 9853, pp. 36–40. Springer, Cham (2016). https://doi.org/10.1007/978-3-319-46131-1_8
5. Fournier-Viger, P., Yang, Y., Yang, P., Lin, J.C.-W., Yun, U.: TKE: mining top-K frequent episodes. In: Fujita, H., Fournier-Viger, P., Ali, M., Sasaki, J. (eds.) IEA/AIE 2020. LNCS (LNAI), vol. 12144, pp. 832–845. Springer, Cham (2020). https://doi.org/10.1007/978-3-030-55789-8_71
6. Huang, K., Chang, C.: Efficient mining of frequent episodes from complex sequences. Inf. Syst. **33**(1), 96–114 (2008)
7. Mannila, H., Toivonen, H., Verkamo, A.I.: Discovering frequent episodes in sequences. In: Proceedings of the 1st International Conference on Knowledge Discovery and Data Mining (1995)
8. Ouarem, O., Nouioua, F., Fournier-Viger, P.: Mining episode rules from event sequences under non-overlapping frequency. In: Proceedings of 34th International Conference on Industrial, Engineering and Other Applications of Applied Intelligent Systems (2021)
9. You, T., Li, Y., Sun, B., Du, C.: Multi-source data stream online frequent episode mining. IEEE Access **8**, 107465–107478 (2020)

Boosting Latent Inference of Resident Preference from Electricity Usage - A Demonstration on Online Advertisement Strategies

Lo Pang-Yun Ting[✉], Po-Hui Wu, Jhe-Yun Jhang, Kai-Jun Yang,
Yen-Ju Chen, and Kun-Ta Chuang

Department of Computer Science and Information Engineering, National Cheng
Kung University, Tainan City, Taiwan
{lpyting,phwu,jyjhang,kjyang,yjchen}@netdb.csie.ncku.edu.tw,
ktchuang@mail.ncku.edu.tw

Abstract. The electricity demand has increased due to the rapid development of the digital economy. The mechanisms of demand-side management are thus proposed to reduce the consumption while electricity companies forecast the appearance of excessive peak load which may incur the instability of electrical grids. However, DSM solutions are generally devised as the way of compulsively controlling home appliances but the interruption is not a pleasurable mechanism. To address this issue, we figure out an advertising strategy based on the residential electricity data acquired from smart meters. By recommending preference-related coupons to residents, we can induce residents to go outside to use the coupon while helping the peak load reduction with pleasure, leading to the win-win result between users and electricity companies. In this paper, we propose a novel framework, called *DMAR*, which combines the *directed inference* and the *mediated inference* to infer residents' preferences based on their electricity usage. Through experimental studies on the real data of smart meters from the power company, we demonstrate that our method can outperform other baselines in the preference inference task. Meanwhile, we also build a line bot system to implement our advertisement service for the real-world residents. Both offline and online experiments show the practicability of the proposed *DMAR* framework.

1 Introduction

In order to pursue the goal of energy saving while preserving the comfortable living need, the demand of electricity surges worldwide, leading to the emergent need of DSM (demand-side management) technologies. Unfortunately, DSM mechanisms are not generally available in most of households due to the high initial cost. In addition, the incentive for users to join the DR program[1] is not

---

[1] Users may generally gain the extra reward according to their reduction amount.

© Springer Nature Switzerland AG 2021
M. Golfarelli et al. (Eds.): DaWaK 2021, LNCS 12925, pp. 272–279, 2021.
https://doi.org/10.1007/978-3-030-86534-4_27

enough to strike the best compromise for users allowing the interruption of their comfortable life. How to achieve the goal of demand-side management still calls for the development of effective strategies for users being willing to reduce their electricity consumption.

In this paper, we propose a new solution to extract the user behavior and demonstrate how to further identify their interests for online advertisement of sending discount e-coupons. The goal is to make a win-win strategy that users would like to go outside for using the desired coupon and electricity company pay similar cost to accomplish the task of demand side management. For example, the coffee shop can provide coupons to attract customers visiting during their off-peak hours. At the same time, the electricity company may forecast the stressful emergent peak load of electricity consumption and would like to execute the DR program to reduce the peak load. While the electricity company can precisely know whether a user, who is using air conditioner now, would like to take a coffee with discount, the coupon can send to the user via online bot or APP to motivate the user going outside and certainly turn off the air conditioner. This business scenario can be beneficial to the electricity company, the coffee shop, and users.

However, it is a challenging issue to infer residents' preferences from their electricity usage data. To resolve this issue, we proposed a novel framework, called *DMAR*, to explore the correlation between electricity behaviors and preferences. In the *DMAR* framework, we consider two kinds of inference approaches to figure out a resident's preference scores for products. The first one is called *Directed Inference*. We assume that different product preferences can be partially inferred by a resident's specific electricity behaviors. The second way is the *Mediated Inference*, which basically considers that a group of residents with similar daily routines may be interested in the same product. The objective of *DMAR* is to infer a resident's preference scores for products, so that we can recommend advertisements which a resident may be most interested in and may want to go outside to purchase products.

One of the goals of existing online advertisement researches is the click-through rate (CTR) prediction for Ads. The researches [2,4,6] predict CTR by adopting the attention mechanism to capture user temporal interests from history behavior sequence. The other goal of the online advertisement is to predict user interest and propose the recommender system [1,3]. However, most of the online advertising mechanisms construct the advertising strategy based on users known interests or their historical interest-related behaviors (e.g., searching records, purchasing records). In our work, we aim to discover a target resident's preference by analyzing his/her electricity usage without knowing a target resident's preference. Therefore, these studies are orthogonal to our work. To the best of our knowledge, this is the first paper to address the mechanism of inferring residents' preferences through electricity usage data.

The structure of this paper is summarized as follows. We present the proposed *DMAR* framework in Sect. 2. The experimental results are exhibited in Sects. 3, and 4 concludes this work.

L. P.-Y. Ting et al.

2 The *DMAR* Model

2.1 Problem Definition

Before introducing our framework, we first give the description of the necessary symbols and definitions. First, let $R = \{r_1, r_2, ...\}$ be a set of residents, and $P = \{p_1, p_2, ...\}$ be a set of products. The electrical feature set of a resident r is defined as $X^r = \{x_{(t_1)}, x_{(t_2)}, ...\}$, where each t_i represents a time period, and each $x_{(t_i)}$ represents an average electrical usage in the i-th time period.

Problem Statement: Given residents R, residents' electrical features $\mathbf{X} = \{X^r | r \in R\}$, and products $P = \{p_1, p_2, ...\}$, our goal is to infer preference scores of products for target residents. We aim to infer the scores $\mathbf{S} = \{S^r | r \in R\}$, where each $S^r = \{s_{p_1}^r, s_{p_2}^r, ...\}$ represents preference scores of all products of a resident r. According to the preference scores, our advertising mechanism can suggest product-related coupons (such as advertisements of the first ranked product) which may strike a chord with the target user and motivate the user moving outside along with turning off electrical appliances.

2.2 Framework

In this section, we introduce the structure of our advertising mechanism and present the *DMAR* (Directed-Mediated Advertising Recommendation) framework to infer residents' preference scores of products from the electricity usage data. The *DMAR* framework combines the *directed inference* and the *mediated inference*. In the *directed inference*, we aim to find the direct correlations between residents' specific electricity behaviors and their preferences, while we put emphasis on discovering the similarity between residents' daily electricity behaviors and inferring preferences based on the group interest in the *mediated inference*. Finally, we use an adaptive association method to combine two inference methods.

The *Directed Inference*. According to the previous researches [5], there exist some correlations between product preferences and specific lifestyles. Hence, in the *direct inference*, we expect that some electricity behaviors in specific time periods have a direct influence on product preferences. We perform a one-way ANOVA F-test to discover electricity features in which time periods could be the most influential to preference inference in the training set. A new feature set F^p for each product p is selected, where each element $\mathbf{f}_r^p \in \mathbb{R}^{1 \times |T^p|}$ in F^p represents selected of electrical features of a resident r for a product p.

Conditional Weighted Regression: In order to model the relationship between residents' preference scores of products and the selected electrical features, we adopt the linear regression technique. Given a target product p, a training feature point $\mathbf{f}_p^r$, and a target feature point $\mathbf{f}_p$, the hypothesis function is defined as $h_{\theta_p}(\mathbf{f}_p^r) = \theta^p \mathbf{f}_p^r + \varepsilon^r$, where θ_p denotes the coefficient vector of the linear equation for each p, and ε^r denotes an error variable. We proposed a *similarity-based weight* and expect that if the similarity of a pair of feature points is bigger than a threshold value which means two feature points are very dissimilar, a distance-decay weight will be assigned to the regression model. The similarity-based weight w_p^r assigned to a pair of feature point $(\mathbf{f}_p^r, \mathbf{f}_p)$ is defined as follow:

$$
w_p^r = \begin{cases} 1 & \text{, if } g(\mathbf{f}_p^r, \mathbf{f}_p) > \varphi \\ \exp\left(- \frac{(\mathbf{f}_p^r - \mathbf{f}_p)^\top (\mathbf{f}_p^r - \mathbf{f}_p)}{2\tau^2} \right) & \text{, otherwise} \end{cases}, \tag{1}
$$

where $g(\cdot)$ represents the cosine similarity measure, φ is the similarity threshold, and τ is the bandwidth parameter. When $g(\mathbf{f}_p^r, \mathbf{f}_p) \le \varphi$, we defined the weight functions as a Gaussian kernel function, which can gain higher values for closer feature points and lower values (tend to zero) for feature points far away. The objective function can be written as: $\min_{\theta,\varepsilon} \sum_{p \in P} \sum_{r \in R} w_p^r \left(h_{\theta_p}(\mathbf{f}_p^r) - \hat{s}_p^r \right)^2$, where $\hat{s}_p^r$ is the real product preference score for an electrical feature point $\mathbf{f}_p^r$, and $\boldsymbol{\theta} = \{\theta_p | p \in P\} \in \mathbb{R}^{1 \times |P|}$. The goal of our regression model is to find the best coefficient vectors $\boldsymbol{\theta}$. Finally, we can gain the inferred product preference scores for a target electrical feature of a resident.

The *Mediated Inference*. The basic assumption of Collaborative Filtering (CF) is that similar people tend to have similar preference. We follow the same observation in the *mediated inference* to assume that those who have similar daily electricity behaviors tend to have the similar product preferences. Therefore, in order to find similar "neighbors" for a target resident, we first divide residents into different groups and infer the product preference scores of a target resident based on the group preference.

Similarity-Based Clustering: To represent the daily electricity behavior of a resident, we average a resident's electricity usages for each day and define a feature vector $\mathbf{x}_r$ as a resident r's average electricity usage in each time period within a day. We design the distance function to represent how similar are residents' electrical behaviors based on two kinds of properties: (1) what time period does a resident usually use electrical appliances, and (2) how much electricity usage does a resident usually use in each time period. Hence, based on the K-Means model, we propose a similarity-based clustering. The objective function at each iteration can be formulated as $\min_M \sum_{\mu_o \in M} \sum_{\mathbf{x}_r \in C_o} \|\mathbf{x}_r - \boldsymbol{\mu}_o\|_D^2$, where $\mathbf{x}_r$ is a feature point of a resident r, and μ_o is its assigned cluster centroids. For each cluster C_o, our method aims to minimize the sum of distances between $\mathbf{x}_r$ and μ_o. In order to consider the two kinds of aforementioned similarity, the proposed distance function $\|\cdot\|_D$ of our similarity-based clustering is designed as the following formula:

$$
\|\mathbf{x}_r - \boldsymbol{\mu}_o\|_D = \frac{\|\mathbf{x}_r - \boldsymbol{\mu}_o\|^2}{\Delta x_{max}^2} (1 - g_n(\mathbf{x}_r, \boldsymbol{\mu}_o)), \tag{2}
$$

where $\|\cdot\|$ is the L^2 norm and Δx_{max} is the maximum value of $\|\mathbf{x}_r - \boldsymbol{\mu}_o\|$ for all pairs of $(\mathbf{x}_r, \boldsymbol{\mu}_o)$. The function $g_n(\cdot)$ is the cosine similarity measure and the value is normalized between 0 to 1. By using the proposed distance function, our clustering method can form a group of residents with similar electricity using times and electricity usages, which means residents who are in the same group have similar daily electricity behaviors. Applying the concept of CF, a target resident's preference scores can be evaluated by averaging scores of all residents who are in the same cluster.

Adaptive Association. After gaining the results of the *directed inference* and the *mediated inference*, we combine two inference methods by adopting a proposed weighted function. The weight ω_D^r is to decide the importance of the *mediated inference* for a resident r, which is based on the dissimilarity between the electricity feature $\mathbf{x}_r$ of a resident r and the assigned group C_o. The weight ω_D^r be formulated as follow:

$$\omega_D^r = \begin{cases} 1 & \text{,if } \|\mathbf{x}_r - \boldsymbol{\mu}_o\|_D \leq \delta \\ 1 - \|\mathbf{x}_r - \boldsymbol{\mu}_o\|_D & \text{,otherwise} \end{cases}, \tag{3}$$

where $\|\cdot\|_D$ is the proposed distance function defined in the Eq. (2), and δ is a dissimilarity threshold value which is defined as $\sum_{\mathbf{x}_{r'} \in C_o} \frac{\|\mathbf{x}_{r'} - \boldsymbol{\mu}_o\|_D}{|C_o|}$ $(r \neq r')$. Hence, we can derive our combined preference scores. The two scores y_{dir}^p, y_{med}^p are defined as a resident's preference scores of a product p inferred by the *direct inference* and the *mediated inference*, respectively. The combined preference score can be derived as $s_p^r = (1 - \omega_D^r)y_{dir}^p + \omega_D^r \cdot y_{med}^p$. Therefore, our proposed framework *DMAR* can get an associated inference scores with an adaptive weight.

3   Experiments

In this section, we conduct extensive experiments to evaluate the effectiveness of the proposed *DMAR* framework. In our experiment, we conduct both offline and online experiments to reveal the effectiveness of *DMAR*. The offline experiment is implemented to answer the first and the second questions, while the online experiment is conducted to answer the third question.

Dataset: We apply a real-world from a power company providing electricity usage data for 39 residents during seven months. We also investigate these residents' preference scores (from 1 to 5) for 15 different kinds of products. These products are classified into 5 categories: *drink, meal, snack, entertainment supply*, and *daily necessity*, which are denoted as C_1, C_2, C_3, C_4, and C_5 respectively. Besides, we conduct a 7-day online experiment on 39 residents. Based on the preference inference results from our *DMAR* framework, we build a LINE bot system which sends advertisements to each resident three times a day. Residents can choose whether they like the recommended products or not in our LINE bot service, and if they like, they can choose whether they would be willing to go outside for purchasing the products.

Comparative Methods: In our experiment, we compare with the following baselines. SBC is the method which only uses the proposed similarity-based clustering; CWR is the method which only utilizes the proposed conditional weighted regression; DM, DM-C, and DM-R are those methods which do not use the adaptive weight for the inference association but with different weights for SBC and CWR ((0.5, 0.5), (0.8, 0.2), and (0.2, 0.8) for DM, DM-C, and DM-R, respectively).

DMAR **Performance:** We compare *DMAR* with aforementioned baselines. The comparison of *DMAR* and baselines are summarized in Table 1. Convincingly, our *DMAR* framework, which aggregates the *directed inference* and the *mediated inference*, and utilizing the proposed adaptive weight for the inference association, can achieve the smallest error value for the preference inference. From the experimental results of SBC and CWR, we can see that only using the directed inference or the mediated inference lead to the worst performance. The results of DM, DM-C, and DM-R, which consider both inference methods without applying the proposed adaptive weight, also are not as good as our method.

Table 1. Comparison of MAE performance.

MAE performance							
Category	SBC	CWR	DM	DM-C	DM-R	DMAR	Improve
C_1	1.4829	1.3351	1.2961	1.3860	1.2796	**0.7745**	[65.2%–91.4%]
C_2	1.2993	1.2501	1.1137	1.2165	1.1605	**0.5282**	[110.8%–145.9%]
C_3	1.2998	1.2607	1.1377	1.2090	1.1956	**0.6529**	[74.2%–99.0%]
C_4	1.1941	1.2375	1.0911	1.1378	1.1581	**0.6341**	[72.0%–95.1%]
C_5	1.1396	1.7778	0.9850	1.0504	1.0707	**0.5559**	[77.1%–219.8%]
Average	1.2832	1.2522	1.1247	1.1999	1.1729	**0.6291**	[78.7%–103.9%]

Online Evaluation: We conduct a 7-day online experiment by implementing the proposed advertisement strategy on our LINE bot system. To prove that our mechanism can really recommend product advertisements in which residents are interested and induce residents to go out, we organize residents' responses after receiving our advertisement recommendation from our LINE bot, which is shown in Fig. 1. In Fig. 1(a), we show the percentages of residents who like our recommended advertisements ('Like') and are willing to go outside to purchase products ('Go out'). We can see that the percentages of 'Like' and 'Go out' are nearly the same in each day. Meanwhile, Fig. 1(b) shows the percentages of 'Not Go Out' and 'Go out' for different category recommendations. It is obvious that the percentage of 'Like' is over 60% and the percentage of 'Go out' is over 30% for each day and for each category. The average percentages of 'Like' and 'Go out' are 0.7559 and 0.4047 in our online experiment, which means that residents are interested in most of our recommendations, and about 40% of residents are willing to go outside to purchase products.

Furthermore, we conduct a simulated experiment to prove that our advertisement strategy has the potential to solve peak load problem. We pick the electricity data and residents' feedback during the 7-day online experiment. Our

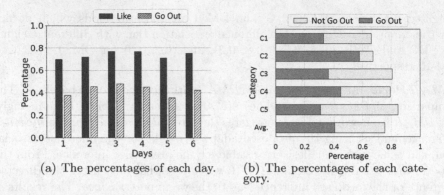

(a) The percentages of each day.   (b) The percentages of each category.

Fig. 1. The results of the online experiment.

scenario is that the LINE bot will send product advertisements to residents before the electricity peak time, and the peak load value will be shifted based on residents' percentages of being willing to go out. Based on the result in Fig. 1, we expect that a resident has 40% probability to go outside at the electricity peak time and will come back home in several hours, which means the electricity usages of residents, who will go outside to purchase products, will be shifted to other time periods. The experimental result shows that our advertisement strategy can reduce about **28.92%** electricity load at the peak time and also keep total electricity usage nearly the same. That is to say, residents do not need to pay extra charges for exceeding the contract capacity, and the electrical company is also able to make the grid system stable.

4   Conclusions

In this paper, we investigate how to infer a resident's preference scores based on electricity behaviors. In the proposed *DMAR* framework, we conduct both the *directed inference* and the *mediated inference* to consider different factors which may affect a resident's preferences. In addition, we define the adaptive association to combine two inference methods based on the dissimilarity measure. Finally, we conduct the offline and online experiments to prove that *DMAR* can outperform other baselines and has the potential to resolve the peak load problem.

References

1. Guo, L., et al.: A deep prediction network for understanding advertiser intent and satisfaction (2020)
2. Liu, B., Tang, R., Chen, Y., Yu, J., Guo, H., Zhang, Y.: Feature generation by convolutional neural network for click-through rate prediction. In: The World Wide Web Conference (2019)
3. Liu, D., Liao, Y.-S., Chung, Y.-H., Chen, K.-Y.: Advertisement recommendation based on personal interests and ad push fairness. Kybernetes **48**, 1586–1605 (2019)

4. Wang, Q., Liu, F., Huang, P., Xing, S., Zhao, X.: A hierarchical attention model for CTR prediction based on user interest. IEEE Syst. J. **14**, 4015–4024 (2020)
5. Westerlund, L., Ray, C., Roos, E.: Associations between sleeping habits and food consumption patterns among 10–11-year-old children in Finland. Br. J. Nutr. **102**, 10:1531–7 (2009)
6. Zhou, G., et al.: Deep interest evolution network for click-through rate prediction. In: AAAI (2019)

Author Index

Printed in the United States
by Baker & Taylor Publisher Services